Inquiries into Truth and Interpretation

DATE DUE			

Inquiries into Truth and Interpretation

DONALD DAVIDSON

CLARENDON PRESS · OXFORD

Oxford University Press, Walton Street, Oxford OX2 6DP
Oxford Glasgow New York Toronto
Delhi Bombay Calcutta Madras Karachi
Kuala Lumpur Singapore Hong Kong Tokyo
Nairobi Dar es Salaam Cape Town
Melbourne Auckland
and associated companies in
Beirut Berlin Ibadan Nicosia

Oxford is a trade mark of Oxford University Press

Published in the United States
by Oxford University Press, New York

Reprinted with corrections 1985

British Library Cataloguing in Publication Data
Davidson, Donald
Inquiries into truth and interpretation.
1. Language—Philosophy
I. Title 401 P106
ISBN 0-19-824617-X
ISBN 0-19-875046-3 Pbk

Library of Congress Cataloging in Publication Data
Davidson, Donald, 1917–
Inquiries into truth and interpretation.
Companion v. to: Essays on actions and events.
Includes bibliographical references and index.
1. Languages—Philosophy—Addresses, essays, lectures. I. Title.
P106.D27 1984 401 83-15136
ISBN 0-19-824617-X
ISBN 0-19-875046-3 (pbk.)

Printed in Great Britain by
J. W. Arrowsmith, Bristol

TO W. V. QUINE

without whom not

Contents

Provenance of the Essays and Acknowledgements

Essay 1, 'Theories of Meaning and Learnable Languages', was read at the 1964 International Congress for Logic, Methodology, and Philosophy of Science at the Hebrew University of Jerusalem. It was subsequently published in the Proceedings of that congress in a volume edited by Yehoshua Bar-Hillel and published by North-Holland Publishing Company, Amsterdam, 1965. It is reprinted here with the permission of the publishers.

An earlier version of Essay 2, 'Truth and Meaning', was read at the Eastern Division meeting of the American Philosophical Association in December 1966. The main theme traces back to a paper delivered at the Pacific Division in 1962. The present essay owes much to John Wallace with whom I discussed these matters from 1962 onward. My research was supported by the National Science Foundation. The paper was first published in *Synthèse*, 17 (1967), 304–23. Copyright © 1967 by D. Reidel Publishing Company, Dordrecht-Holland, and reprinted by permission of D. Reidel Publishing Company.

Essay 3, 'True to the Facts', was first presented at a symposium on Truth in December 1969 at a meeting of the Eastern Division of the American Philosophical Association. The other symposiast was James F. Thomson. The paper first appeared in the *Journal of Philosophy*, 66 (1969), 748–64, and is printed here with the permission of the editors.

'Semantics for Natural Languages', Essay 4, was read at a symposium organized by the Olivetti Company in honour of its founder, and held in Milan in October 1968. The proceedings were published in *Linguaggi nella Società e nella Tecnica*, Edizioni di Comunità, Milan, 1970.

Essay 5, 'In Defence of Convention T', was read at a conference on Alternative Semantics held at Temple University in December 1970, and was published in *Truth, Syntax and Modality* (the title *Truth Valued*, suggested by Dana Scott, was unfortunately rejected). The book was edited by Hugues Leblanc and published by North-Holland Publishing Company, 1973. It is reprinted here by permission of the publishers.

The next Essay, number 6, 'Quotation', was published in a special issue of *Theory and Decision* on Language Theory edited by H. L. Berghel (*Theory and Decision*, 11 (1979), 27–40). Copyright © 1979 by D. Reidel Publishing Company, Dordrecht-Holland. Reprinted by permission of D. Reidel Publishing Company.

'On Saying That', which is Essay 7, was published in a double issue of *Synthèse* devoted to the work of W. V. Quine (*Synthèse*, 19 (1968–9), 130–46). It was subsequently published in *Words and Objections, Essays on the Work of W. V. Quine*, edited by D. Davidson and J. Hintikka, D. Reidel, 1969, pp. 158–74 (revised edition, 1975). Copyright © 1969 by D. Reidel Publishing Company, Dordrecht-Holland. Reprinted by permission of D. Reidel Publishing Company. Quine's reply to this essay is on pages 333–5 of *Words and Objections*.

Essay 8, 'Moods and Performances', was read at the second Jerusalem Philosophical Encounter, held in Israel in April 1976, and was commented on by W. V. Quine. It was published in *Meaning and Use*, edited by A. Margalit, D. Reidel Publishing Company, Dordrecht-Holland, 1979.

'Radical Interpretation', Essay 9, was read at a colloquium on Philosophical Problems of Language in Biel, Switzerland, in May 1973, and in another version at a conference on Language and Meaning at Cumberland Lodge, Great Park, Windsor in November of that year. It was published in *Dialectica*, 27 (1973), 313–28, and is reprinted here with the permission of the editor, H. Lauener.

Essay 10 on 'Belief and the Basis of Meaning' was prepared for a conference on Language, Intentionality, and Translation Theory held at the University of Connecticut in March 1973, and was published in a double issue of *Synthèse* edited by J. G. Troyer and S. C. Wheeler III (*Synthèse*, 27 (1974), 309–23). This issue also contains valuable comments by W. V. Quine (325–9) and David Lewis (331–44) and my replies (345–9). David Lewis's reply has the title 'Radical Interpretation'. Copyright © 1974 by D. Reidel Publishing

Company, Dordrecht-Holland. Reprinted by permission of D. Reidel Publishing Company.

Essay 11, 'Thought and Talk', was a 1974 Wolfson College lecture, and was published in *Mind and Language*, edited by Samuel Guttenplan, © Oxford University Press, 1975. It is reprinted here with the permission of Oxford University Press on behalf of Wolfson College.

In June 1974 John Foster read a paper, 'Meaning and Truth Theory', to the Oxford Philosophical Society, and Essay 12, 'Reply to Foster', is my reply to the parts of his paper that concerned my work. Foster's paper and my response were published in *Truth and Meaning: Essays in Semantics*, edited by Gareth Evans and John McDowell, © Oxford University Press, 1976. My paper is reprinted here with the permission of the Oxford University Press.

Essay 13, 'On the Very Idea of a Conceptual Scheme', was slow to reach its present form. The sixth and last of my John Locke lectures, titled 'Invariants of Translation', was an early draft, delivered in Trinity Term, 1970, at Oxford. In January of the next year I gave two lectures on Alternative Conceptual Schemes at the University of London which contained much of what is in the present essay. The material was distilled down to almost final form for my presidential address to the Eastern Meeting of the American Philosophical Association in Atlanta, 28 December 1973. After that, but before publication, I gave a closely related talk, 'The Third Dogma of Empiricism', to the Philosophical Society at Oxford. The discussion was opened by W. V. Quine, and his comments helped me write the final draft. Some later fall-out from that discussion will be found in Quine's 'On the Very Idea of a Third Dogma'. My paper was published in the *Proceedings and Addresses of the American Philosophical Association*, 47 (1974), and is reprinted here with the permission of the Association.

Essay 14, 'The Method of Truth in Metaphysics', was first published in *Midwest Studies in Philosophy, 2: Studies in the Philosophy of Language*, edited by P. A. French, T. E. Uehling, Jr., and H. K. Wettstein, The University of Minnesota, Morris, 1977. To my great profit, Gilbert Harman and W. V. Quine commented on earlier versions.

Essay 15, 'Reality Without Reference', was first given in different form at a Semantics and Linguistics Workshop at the University of Western Ontario, London, Ontario in April 1972. Criticisms and suggestions followed, and changed, the paper over the years. It was first published in *Dialectica*, 31 (1977), 247–53, and is reprinted here with the permission of the editor.

'The Inscrutability of Reference', Essay 16, was written for a special issue of *The Southwestern Journal of Philosophy* devoted to the work of W. V. Quine, but I missed the deadline and the paper appeared in a later issue, *The Southwestern Journal of Philosophy*, 10 (1979), 7–19. It is reprinted here with permission of the Journal. Quine replied to my paper along with others in 'Replies to the Eleven Essays', *Philosophical Topics*, 11 (1981), 242–3.

Essay 17, 'What Metaphors Mean', was read at a conference on metaphor at the University of Chicago in February 1978. It was first published in *Critical Inquiry*, 5 (1978), 31–47; © 1978 by Donald Davidson. The extract from 'The Hippopotamus' on p. 256 is from *Collected Poems 1909–1962* by T. S. Eliot, copyright 1936 by Faber and Faber and Harcourt Brace Jovanovich, Inc.; copyright © 1963, 1964 by T. S. Eliot. It is reprinted by permission of the publishers. In a subsequent issue of the same journal, Nelson Goodman and Max Black responded to my piece. Neither found much to agree with in what I had written. (Nelson Goodman, 'Metaphor as Moonlighting', *Critical Inquiry*, 6 (1979), 125–30 and Max Black, 'How Metaphors Work: A Reply to Donald Davidson', *Critical Inquiry*, 6 (1979), 131–43.)

The last essay, Essay 18, on 'Communication and Convention', was read at the first Campinas Encounter in the Philosophy of Language at the Universidade Estadual de Campinas in August 1981. It is scheduled to be published in the proceedings of the Encounter, *Dialogue: an Interdisciplinary Approach*, edited by Marcelo Dascal, John Benjamins, Amsterdam.

Introduction

What is it for words to mean what they do? In the essays collected here I explore the idea that we would have an answer to this question if we knew how to construct a theory satisfying two demands: it would provide an interpretation of all utterances, actual and potential, of a speaker or group of speakers; and it would be verifiable without knowledge of the detailed propositional attitudes of the speaker. The first condition acknowledges the holistic nature of linguistic understanding. The second condition aims to prevent smuggling into the foundations of the theory concepts too closely allied to the concept of meaning. A theory that does not satisfy both conditions cannot be said to answer our opening question in a philosophically instructive way.

The first five essays are mainly concerned with the question what sort of a theory would satisfy the first condition.

Essay 1, 'Theories of Meaning and Learnable Languages', urges that a satisfactory theory must discover a finite basic vocabulary in the verbal phenomena to be interpreted if it is to prove useful to a creature with finite powers. If this is so, there is no escape from the need to treat the semantic features of the potential infinity of sentences as owed to the semantic features of the items in a finite vocabulary. It turns out that a number of familiar theories fail to meet this condition: Frege's analysis of oblique contexts, Church's logic of sense and denotation, Tarski's informal treatment of quotation are examples. Standard theories of adverbial modification might well be added to the list.

Essay 2, 'Truth and Meaning', argues that a theory of truth along the lines of Tarski's truth definitions, but modified in various ways to apply to a natural language, would be enough for an interpreter to go on. Such theories have clear virtues. They make no use of meanings as entities; no objects are introduced to correspond to predicates or sentences; and from a finite set of axioms it is possible to prove, for each sentence of the language to be interpreted, a theorem that states truth conditions for that sentence. Further, the proof of such a theorem amounts to an analysis of how the truth or falsity of the sentence depends on how it is composed from elements drawn from the basic vocabulary. If such theories really do satisfy the two conditions listed in the first paragraph, we can take the word 'theory' in 'theory of meaning' seriously.

Many objections have been made to the claim that truth theories can do duty as theories of meaning. Some of the objections I have tried to meet or deflect in other essays in this book. But whether or not the claim can be made good, some of the arguments for it in 'Truth and Meaning' are faulty. The reader will find that I shifted ground more than once as I tried to improve or clarify this central thesis. One thing that only gradually dawned on me was that while Tarski intended to analyse the concept of truth by appealing (in Convention T) to the concept of meaning (in the guise of sameness of meaning, or translation), I have the reverse in mind. I considered truth to be the central primitive concept, and hoped, by detailing truth's structure, to get at meaning. These are remarks *about* theories of truth, of course, not remarks to be found in them.

Something else that was slow coming to me was that since I was treating theories of truth as empirical theories, the axioms and theorems had to be viewed as laws. So a theorem like '"Schnee ist weiss" is true in the mouth of a German speaker if and only if snow is white' has to be taken not merely as true, but as capable of supporting counterfactual claims. Indeed, given that the evidence for this law, if it is one, depends ultimately on certain causal relations between speakers and the world, one can say that it is no accident that 'Schnee ist weiss' is true if and only if snow is white; it is the whiteness of snow that *makes* 'Schnee ist weiss' true. How much of a concession this is to intensionality depends, I suppose, on one's analysis of the concept of law. What seems clear is that whatever the concession comes to, it is one that must be made for any empirical science. These matters are discussed in Essay 12.

Essay 3, 'True to the Facts', asks whether a theory of truth in Tarski's style should be called a correspondence theory. Such theories do not, like most correspondence theories, explain truth by finding entities such as facts for true sentences to correspond to. And there are good reasons, which can be traced back to Frege, for rejecting facts as entities that could play this role. On the other hand, theories of truth of the kind considered here do require that a relation between entities and expressions be characterized ('satisfaction'). It is not easy to see how a satisfactory route to truth can escape this step if the language the theory treats has the usual quantificational resources.

'Semantics for Natural Languages', *Essay 4*, urges that truth theories could provide a formal semantics for natural languages to match the sort of formal syntax linguists from Chomsky on have favoured. When this essay was written, the deep structures of syntax were thought to be the vehicles for semantic interpretation. Essay 4 suggested that the deep structure of a sentence should correspond to the logical form a theory of truth assigned to that sentence.

Tarski's Convention T, which is defended in *Essay 5*, is an informal, but powerful, instrument for testing theories of truth against one's prior grasp of the concept. In the most direct application, the test merely calls on us to recognize the disquotational feature of truth predicates; sentences like '"Snow is white" is true in English if and only if snow is white' are trivially true. Since the totality of such sentences uniquely determines the extension of a truth predicate for English, a theory that entails all such sentences must be extensionally correct. Critics have often made the error of thinking that since the theorems that show a theory to be correct are trivial, the theory or the concept of truth it characterizes, must also be trivial.

A theory of truth would serve to interpret a speaker only if the theory were up to accounting for all the linguistic resources of the speaker. But is a theory that satisfies Convention T adequate to a natural language? Here there are two questions. One is what devices to make or consider available in the language of the theory; the other is how to apply these devices to the language of the speaker. My working assumption has been that nothing more than standard first-order quantification theory is available. Indeed, I was long convinced that many alternative approaches to semantics, employing, for example, modal logics, possible world semantics, or substitutional quantification, could not be accommodated in a theory that met the demands of Convention T. I now know this was hasty. Convention T does not settle as much as I thought, and more possibilities for interesting theorizing are open than I had realized. The well-known virtues of first-order quantification theory still provide plenty of motivation, however, to see how much we can do with it. In the next three essays, collected under the head of application, I attempt the semantic taming of three related but recalcitrant idioms: quotation, indirect discourse, and mood operators.

Essay 6 points out that no current theory of quotation is entirely satisfactory, and it proposes an explicitly demonstrative approach which makes quotation a special case of the demonstrative reference of words to other words in the verbal neighbourhood.

Essay 7, 'On Saying That', concentrates on one of the many kinds of sentence used to attribute attitudes; the paratactic solution suggested has obvious affinities with the treatment of quotation in Essay 6. In Essay 3 there are hints (which I think could be developed) on how the analysis could be extended to belief sentences. If the strategy were to be pursued, it might serve to give a semantics (though not a logic) for the modalities, for counterfactuals, and further sentences about 'propositional' attitudes.

Essay 8, 'Moods and Performances', stresses the often neglected distinction between grammatical moods on the one hand and various sorts of illocutionary force on the other. Only the first is of concern to a theory of what words mean. Here a paratactic analysis of imperatives is suggested which is intended to accommodate our natural feeling that imperatives don't have a truth value while remaining within the resources of a theory of truth.

In the companion volume to this one, *Essays on Actions and Events*, I show how a theory of truth can be applied to a number of further problem cases: sentences about actions and other events, adverbial modification, and singular causal statements.

The third section of the present book is addressed to the question whether a theory of truth for a speaker can be verified without assuming too much of what it sets out to describe.

In 'Radical Interpretation', *Essay 9*, as in the rest of the essays, I follow Quine in supposing that even if we narrow attention to verbal behaviour that reveals when, and under what conditions, a speaker gives credence to a sentence, there is no direct way of sorting out the roles of belief and meaning in explaining that credence. Eliciting separate accounts of belief and meaning requires a theory that can bring to bear on the interpretation of each sentence and its accompanying attitudes the contribution of further data. Only by studying the *pattern* of assents to sentences can we decide what is meant and what believed.

Depending on evidence which, without the aid of theory, makes no distinction between the contributions of belief and meaning to linguistic behaviour, requires a method for effecting the separation to a degree sufficient for communication. Devices to this end are described and defended in the present essays. But all of them, in one way or another, rely on the Principle of Charity.

The phrase and the basic idea come from Neil Wilson, 'Substances Without Substrata'. Quine puts it this way: '. . . assertions startlingly false on the face of them are likely to turn on hidden differences of language' (*Word and Object*, p. 59). Quine applies the principle primarily to the interpretation of the logical constants.

Because I find I cannot use Quine's notion of stimulus meaning as a basis for interpreting certain sentences, I apply the Principle of Charity across the board. So applied, it counsels us quite generally to prefer theories of interpretation that minimize disagreement. So I tended to put the matter in the early essays, wanting to stress the inevitability of the appeal to charity. But minimizing disagreement, or maximizing agreement, is a confused ideal. The aim of interpretation is not agreement but understanding. My point has always been that understanding can be secured only by interpreting in a way that makes for the right sort of agreement. The 'right sort', however, is no easier to specify than to say what constitutes a good reason for holding a particular belief.

The subtle pressures on the Principle of Charity begin to emerge in Essays 10 and 11. Yet here too there are only hints; in work now in progress I attempt to develop the subject in more detail.

Essay 10, 'Belief and the Basis of Meaning', insists on the symmetry of belief and meaning in the exploration of verbal behaviour. In one important respect it goes further. It develops a striking parallel between Bayesian theories of decision and theories of meaning, and gives reasons why the two theories should be considered mutually dependent. The hints dropped here, which give promise of a unified theory of speech and action, have been taken up in my Carus Lectures, and will be published presently.

The first two essays on radical interpretation stress the fact that understanding the words of a speaker requires knowing much about what he believes. *Essay 11*, 'Thought and Talk', attends to the reciprocal dependence, and concludes, rather speculatively, that only a creature with a language can properly be said to have a full-fledged scheme of propositional attitudes.

Essay 12, 'Reply to Foster', as remarked above, recognizes that if a theory of truth is to suffice for interpretation, it must be more than true: its axioms and theorems must be natural laws. If an interpreter knew such a theory, he could use it to understand a speaker, but only if he knew that the theory's pronouncements were nomic.

The next four essays may be described as philosophical fall-out from the approach to truth and interpretation recommended here.

A theory of truth can be called a correspondence theory in the unassuming sense of Essay 3, but that sense does not encourage the thought that we understand what it would be like to compare sentences with what they are about, since the theory provides no entities with which to compare sentences. Along related lines, *Essay 13*, 'On the Very Idea of a Conceptual Scheme', scouts the intelligibility of claims that different languages or conceptual schemes 'divide up' or 'cope with' reality in importantly different ways. Our general method of interpretation forestalls the possibility of discovering that others have radically different intellectual equipment. But more important, it is argued that if we reject the idea of an uninterpreted source of evidence no room is left for a dualism of scheme and content. Without such a dualism we cannot make sense of conceptual relativism. This does not mean that we must give up the idea of an objective world independent of our knowledge of it. The argument against conceptual relativism shows rather that language is not a screen or filter through which our knowledge of the world must pass.

Giving up the dualism of scheme and content amounts to abandoning a theme central to empiricism in its main historical manifestations. But I do not think, as friends and critics have variously suggested, that my argument against empiricism makes me, or ought to make me, a pragmatist, a transcendental idealist, or an 'internal' realist. All these positions are forms of relativism that I find as hard to understand as the empiricisms I attack.

According to Essay 13, no sense can be made of the idea that the conceptual resources of different languages differ dramatically. The argument that makes for this conclusion makes equally for the conclusion that the general outlines of our view of the world are correct; we individually and communally may go plenty wrong, but only on condition that in most large respects we are right. It follows that when we study what our language—any language—requires in the way of overall ontology, we are not just making a tour of our own picture of things: what we take there to be is pretty much what there is. This is the theme of *Essay 14*, 'The Method of Truth in Metaphysics'.

A theory of truth is tested by theorems that state the conditions under which sentences are true; these theorems say nothing about reference. *Essay 15*, 'Reality Without Reference', accordingly contends that how a theory of truth maps non-sentential expressions on to objects is a matter of indifference as long as the conditions of truth are not affected. The question what objects a particular sentence is about, like the questions what object a term refers to, or what objects a predicate is true of, has no answer.

In Essay 15 I am with Quine in holding reference to be inscrutable. *Essay 16*, 'The Inscrutability of Reference', warns against taking inscrutability as a reason for trying somehow to relativize the reference and ontology of singular terms and predicates. For since nothing can reveal how a speaker's words have been mapped on to objects, there is nothing to relativize to; and interpretation being unaffected, there is no need to relativize.

No discussion of theories of meaning can fail to take account of the limits of application of such theories. The scope must be broad enough to provide an insight into how language can serve our endless purposes, but restricted enough to be amenable to serious systematization. Essay 8 took a necessary step by distinguishing between grammatical mood, which the meanest theory must account for, and the force of utterances, which is beyond the reach of comparable regimentation. *Essay 17*, 'What Metaphors Mean', is mainly devoted to the thesis that we explain what words in metaphor do only by supposing they have the same meanings they do in non-figurative contexts. We lose our ability to account for metaphor, as well as rule out all hope of responsible theory, if we posit metaphorical meanings.

Essay 18, 'Communication and Convention', draws another boundary. It is always an open question how well the theory an interpreter brings to a linguistic encounter will cope. In practice an interpreter keeps the conversation going by adjusting his theory on the spot. The principles of such inventive accommodation are not themselves reducible to theory, involving as they do nothing less than all our skills at theory construction.

The essays have been retouched in minor ways to reduce repetition, to eliminate unnecessary or confused passages, or to bring early more into line with later thoughts. These temperings have been limited to the trivial. Where my errors or lapses have earned attention I have let things stand, or marked the change with a footnote.

Many more people have helped me than I can possibly thank here, but I do especially want to mention Paul Grice, Gilbert Harman, Saul Kripke, David Lewis, Richard Rorty, Sir Peter Strawson, and Bruce Vermazen. Sue Larson and Akeel Bilgrami did indispensable work on the footnotes, bibliography, and index. Much more than that, they gave me philosophical advice and moral support. Sue Larson has taught me much about philosophy of language; her influence is especially strong in Essays 8 and 18.

In 1970 I gave the John Locke lectures at Oxford. The contents of those lectures turn up here (much modified) in Essays 2, 3, 6, 7, and 13. A further lecture on adverbial modification drew on material now printed in Essays 6–11 of *Essays on Actions and Events*.

An early influence on my thinking was Michael Dummett, who lectured on Frege and philosophy of language several times at Stanford University while I was there in the fifties. Our discussions took a public form in 1974 when we gave a joint seminar on truth while I was a visiting fellow at All Souls College.

Over the years John Wallace and I talked endlessly about the issues raised in this book. He early appreciated the power of Tarski's work on truth, and much that I have written reflects his insight and sympathetic criticism.

W. V. Quine was my teacher at a crucial stage in my life. He not only started me thinking about language, but he was the first to give me the idea that there is such a thing as being right, or at least wrong, in philosophy, and that it matters which. Without the inspiration of his writing, his patient tutelage, his friendly wit and his generous encouragement, this book would not be worse than it is. It would not be.

TRUTH AND MEANING

1 *Theories of Meaning and Learnable Languages*

Philosophers are fond of making claims concerning the properties a language must have if it is to be, even in principle, learnable. The point of these claims has generally been to bolster or to undermine some philosophical doctrine, epistemological, metaphysical, ontological, or ethical. But if the arguments are good they must have implications for the empirical science of concept formation, if only by way of saying what the limits of the empirical are.

Often it is asserted or implied that purely a priori considerations suffice to determine features of the mechanisms, or the stages, of language learning; such claims are suspect. In the first part of this paper I examine a typical example of such a position, and try my hand at sorting out what may be acceptable from what is not. In contrast to shaky hunches about how we learn language, I propose what seems to me clearly to be a necessary feature of a learnable language: it must be possible to give a constructive account of the meaning of the sentences in the language. Such an account I call a theory of meaning for the language, and I suggest that a theory of meaning that conflicts with this condition, whether put forward by philosopher, linguist, or psychologist, cannot be a theory of a natural language; and if it ignores this condition, it fails to deal with something central to the concept of a language. Nevertheless, as I try to show in the second part of this paper, a number of current theories of meaning do either conflict with or ignore this condition for being learnable.

I

First we learn a few names and predicates that apply to medium-sized lovable or edible physical objects in the foreground of sense and interest; the learning takes place through a conditioning process involving ostension. Next come complex predicates and singular terms for objects not necessarily yet observed, or forever out of sight due to size, date, attenuation, or inexistence. Then come theoretical terms, learned perhaps by way of 'meaning postulates' or by dint of being embedded in suitably scientific discourse. Somewhere early in the game the great jump is made from term to sentence, though just how may be obscure, the transition being blurred by the existence of one-word sentences: 'Mama', 'Fire', 'Slab', 'Block', 'Gavagai', and so on.

Thus, in brief caricature, goes the building-block theory of language learning, echoing, chapter by dusty chapter, empiricist epistemology.

The theory is now discredited in most details. For one thing, there is no obvious reason to think the order of learning is related to epistemological priority. For another, some of the claims seem contradicted by experience: for example, a child learns the general terms 'cat', 'camel', 'mastodon', and 'unicorn' in what may be, to all telling, a uniform way (perhaps by paging through a picture book), though the child's relation to the extensions of these terms is altogether different. In some cases, the order of language learning is arguably the reverse of the epistemological order: sense-data may be the basis for our knowledge of physical objects, but talk of sense-data is learned, if learned at all, long after talk of physical objects is achieved. Finally, the underlying epistemology, with its assumption of associationist psychology and its simple reductionist theory of meaning is no longer appealing to most philosophers. In the light of all this it is astonishing that something very like the doctrine of language learning which began as a feeble outgrowth of early empiricism should now flourish while the parent plant wilts. What follows is a single example of dependence on this outmoded doctrine, but the example could easily be multiplied from current literature.

P. F. Strawson has attacked Quine's well known view that 'the whole category of singular terms is theoretically superfluous'.[1] Strawson grants, at least for argument's sake, that within a language already containing singular terms, we can paraphrase 'all that we at

[1] P. F. Strawson, 'Singular Terms, Ontology and Identity'. Subsequent references give the page numbers in parentheses. The quotation from Quine is on page 211 of *Methods of Logic*.

present say with the use of singular terms into forms of words which do not contain singular terms' (434). What Strawson denies is that from this assumption there follows the theoretical possibility that we could speak a language without singular terms, '... in which we never had used them, in which the category of singular terms simply did not exist, but in which we were nevertheless able to say, in effect, all that we are at present able to say with the use of singular terms' (433, 434). Strawson then sets out to establish independently the theoretical impossibility of such a language.

In order to focus on the point at issue, let me explain that I have no interest in contesting two of Strawson's theses, namely, and roughly, the *eliminability* of singular terms does not follow from *paraphrasability*, and that eliminability, as described, is impossible. Both points are moot, so far as I am concerned, pending clarification of the notion of being able to say the same thing.

Not to try conclusions, then, my concern is entirely with the main argument Strawson uses in the attempt to discredit Quine's eliminability thesis. Two claims essential to this argument are these:

(1) for any predicate to be understood, some predicates must be learned ostensively or by 'direct confrontation';

(2) for such learning to take place, the ostensive learning situation must be 'articulated in the language' by a demonstrative element which picks out or identifies entities of the sort to which the predicate applies (445, 446).

Quine has countered that (1) and (2) do not suffice to establish the necessity for singular terms, because demonstratives may be construed as general terms.[2] This may well be true; my interest however is in the implication of (1) and (2) that substantive questions about language learning can be settled on purely a priori grounds.[3]

Summarizing his own argument, Strawson says: 'Some universal terms must be connected with our experience if any are to be understood. And these universal terms must be connected with particular bits or slices of our experience. Hence, if they are to be

[2] W. V. Quine, *Word and Object*, 185.

[3] Strawson quotes Quine, *Mathematical Logic*, 218, in support of (1), and Quine himself might now be quoted in support of something like both (1) and (2) in *Word and Object* and *The Roots of Reference*. But aside from the fact that Quine declines to draw Strawson's conclusion, there remains the important difference that Quine at most thinks (1) and (2) are true, while Strawson wants to show they are necessary. I am concerned here only with the claim of necessity, though I doubt that either (1) or (2) is an important truth about language acquisition.

learnt *as predicates of particulars*, they must be learnt as predicates of demonstratively *identified* particulars.' (446. Italics in original.) Here it is perhaps obvious that the notion of learning appears vacuously in the conclusion; so let us turn briefly to (1) and (2).

Surely it is an empirical question whether, as a result of certain experiences, a person turns out to have some ability he did not have before; yet (1) and (2) claim it is a purely 'logical' matter that everyone who has acquired a linguistic ability of a specified kind has travelled a prescribed route. Strawson apparently equates the ostensive learning of a predicate with learning by 'direct confrontation'. One can imagine two ways in which such a process is intended to be more special than learning the meaning of a predicate through hearing sentences which couple it with demonstrative singular terms. One is, that ostensive learning may require an intention on the part of the teacher to bring an object to the attention of the learner. It seems however that no such intention is necessary, and in fact most language learning is probably due more to observation and imitation on the part of the learner than to any didactic purpose on the part of those observed and imitated. A second difference is that direct confrontation (and probably ostension, at least as Strawson interprets it) requires the presence of an appropriate object, while the correct use of a demonstrative singular term does not. But it is not an a priori truth, nor probably even a truth, that a person could not learn his first language in a skillfully faked environment. To think otherwise would, as Strawson says in a later book, but perhaps in much the same connection, 'be to limit too much the power of human imagination'.[4]

In defending (2) Strawson reveals what I think is the confusion that underlies the argument I have been criticizing. On the face of it, there is no reason why the ostensive learning of predicates must be 'articulated in the language' in one way rather than another; no reason why the reference to particulars (assuming this necessary) must be by way of demonstratives. Nor does Strawson provide any reason; instead he gives grounds for the claim that 'no symbolism can be interpreted as a language in which reference is made to particulars, unless it contains devices for making demonstrative . . . references to particulars, i.e. unless it contains singular terms for referring to particulars . . .'.[5] Later in the same paper Strawson

[4] P. F. Strawson, *Individuals*, 200.
[5] P. F. Strawson, 'Singular Terms, Ontology and Identity', 447.

attacks '. . . the uncritical assumption that a part only of the structure of ordinary language could exist and function in isolation from the whole of which it is a part, just as it functions when incorporated within that whole . . . To this extent, at least, language is organic' (451, 452). Here the thought emerges, clear of any important connections with language learning, that Quine's eliminability thesis is false because we would make a major conceptual change—alter the meanings of all retained sentences—if we cut them off from their present relations with sentences containing singular terms (or demonstratives, or proper names, etc.). I suggest that from arguments for the conceptual interdependence of various basic idioms, Strawson has illegitimately drawn conclusions concerning the mechanism and sequence of language acquisition.

On the question of conceptual interdependence, Quine seems, up to a point at least, in agreement, writing that 'the general term and the demonstrative singular are, along with identity, interdependent devices that the child of our culture must master all in one mad scramble'.[6] It remains true, if such claims of interdependence are tenable, that we could not learn a language in which the predicates meant what predicates in our language do mean and in which there were no demonstrative singular terms. We could not learn such a language because, granting the assumption, there could not be such a language.[7] The lesson for theories of language learning is wholly negative, but not perhaps without importance: in so far as we take the 'organic' character of language seriously, we cannot accurately describe the first steps towards its conquest as learning part of the language; rather it is a matter of partly learning.[8]

II

It is not appropriate to expect logical considerations to dictate the route or mechanism of language acquisition, but we are entitled to consider in advance of empirical study what we shall count as knowing a language, how we shall describe the skill or ability of a

[6] W. V. Quine, *Word and Object*, 102.

[7] In *Individuals* Strawson again attacks what he takes to be Quine's eliminability thesis, but does not use the argument from learning which I am criticizing.

[8] This theme is taken up again in Essay 11. In Essay 8 of *Essays on Action and Events* I discuss the question whether objects are conceptually prior to events.

person who has learned to speak a language. One natural condition to impose is that we must be able to define a predicate of expressions, based solely on their formal properties, that picks out the class of meaningful expressions (sentences), on the assumption that various psychological variables are held constant. This predicate gives the grammar of the language. Another, and more interesting, condition is that we must be able to specify, in a way that depends effectively and solely on formal considerations, what every sentence means. With the right psychological trappings, our theory should equip us to say, for an arbitrary sentence, what a speaker of the language means by that sentence (or takes it to mean). Guided by an adequate theory, we see how the actions and dispositions of speakers induce on the sentences of the language a semantic structure. Though no doubt relativized to times, places, and circumstances, the kind of structure required seems either identical with or closely related to the kind given by a definition of truth along the lines first expounded by Tarski, for such a definition provides an effective method for determining what every sentence means (i.e. gives the conditions under which it is true).[9] I do not mean to argue here that it is necessary that we be able to extract a truth definition from an adequate theory (though something much like this is needed), but a theory meets the condition I have in mind if we can extract a truth definition; in particular, no stronger notion of meaning is called for.[10]

These matters appear to be connected in the following informal way with the possibility of learning a language. When we can regard the meaning of each sentence as a function of a finite number of features of the sentence, we have an insight not only into what there is to be learned; we also understand how an infinite aptitude can be encompassed by finite accomplishments. For suppose that a language lacks this feature; then no matter how many sentences a would-be speaker learns to produce and understand, there will remain others whose meanings are not given by the rules already mastered. It is natural to say such a language is *unlearnable*. This argument depends, of course, on a number of empirical assump-

[9] A. Tarski, 'The Concept of Truth in Formalized Languages'.

[10] Many people, including Tarski, have thought it was impossible to give a truth definition even for the indicative sentences of a natural language. If this is so, we must find another way of showing how the meanings of sentences depend on their structures. See Essays 2, 4, and 9.

tions: for example, that we do not at some point suddenly acquire an ability to intuit the meanings of sentences on no rule at all; that each new item of vocabulary, or new grammatical rule, takes some finite time to be learned; that man is mortal.

Let us call an expression a *semantical primitive* provided the rules which give the meaning for the sentences in which it does not appear do not suffice to determine the meaning of the sentences in which it does appear. Then we may state the condition under discussion by saying: a learnable language has a finite number of semantical primitives. Rough as this statement of the condition is, I think it is clear enough to support the claim that a number of recent theories of meaning are not, even in principle, applicable to natural languages, for the languages to which they apply are not learnable in the sense described. I turn now to examples.

First example. Quotation Marks. We ought to be far more puzzled than we are by quotation marks. We understand quotation marks very well, at least in this, that we always know the reference of a quotation. Since there are infinitely many quotations, our knowledge apparently enshrines a rule. The puzzle comes when we try to express this rule as a fragment of a theory of meaning.

Informal stabs at what it is that we understand are easy to come by. Quine says, 'The name of a name or other expression is commonly formed by putting the named expression in single quotation marks ... the whole, called a *quotation*, denotes its interior';[11] Tarski says essentially the same thing.[12] Such formulas obviously do not provide even the kernel of a theory in the required sense, as both authors are at pains to point out. Quine remarks that quotations have a 'certain anomalous feature' that 'calls for special caution'; Church calls the device 'misleading'. What is misleading is, perhaps, that we are tempted to regard each matched pair of quotation marks as a functional expression because when it is clamped around an expression the result denotes that expression. But to carry this idea out, we must treat the expressions inside the quotation marks as singular terms or variables. Tarski shows that in favourable cases, paradoxes result; in unfavourable cases, the expression within has no meaning (159–62). We must therefore give up the idea that quotations are 'syntactically composite expressions,

[11] W. V. Quine, *Mathematical Logic*, §4.
[12] A. Tarski, 'The Concept of Truth in Formalized Languages', 156.

of which both the quotation marks and the expressions within them are parts'. The only alternative Tarski offers us is this:

> Quotation-mark names may be treated like single words of a language and thus like syntactically simple expressions. The single constituents of these names—the quotations marks and the expressions standing between them—fulfil the same function as the letters and complexes of successive letters in single words. Hence they can possess no independent meaning. Every quotation-mark name is then a constant individual name of a definite expression . . . and in fact a name of the name nature as the proper name of a man (159).

In apparently the same vein, Quine writes that an expression in quotation marks 'occurs there merely as a fragment of a longer name which contains, besides this fragment, the two quotation marks', and he compares the occurrence of an expression inside quotation marks with the occurrence of 'cat' in 'cattle'[13] and of 'can' in 'canary'.[14]

The function of letters in words, like the function of 'cat' in 'cattle', is purely adventitious in this sense: we could substitute a novel piece of typography everywhere in the language for 'cattle' and nothing in the semantical structure of the language would be changed. Not only does 'cat' in 'cattle' not have a 'separate meaning'; the fact that the same letters occur together in the same order elsewhere is irrelevant to questions of meaning. If an analogous remark is true of quotations, then there is no justification in theory for the classification (it is only an accident that quotations share a common feature in their spelling), and there is no significance in the fact that a quotation names 'its interior'. Finally, every quotation is a semantical primitive, and, since there are infinitely many different quotations, a language containing quotations is unlearnable.

This conclusion goes against our intuitions. There is no problem in framing a general rule for identifying quotations on the basis of form (any expression framed by quotation marks), and no problem in giving an informal rule for producing a wanted quotation (enclose the expression you want to mention in quotation marks). Since these rules imply that quotations have significant structure, it is hard to deny that there must be a semantical theory that exploits it. Nor is it entirely plain that either Tarski or Quine wants to deny the possibility. Tarski considers only two analyses of quotations, but he

[13] W. V. Quine, *From a Logical Point of View*, 140.
[14] W. V. Quine, *Word and Object*, 144.

does not explicitly rule out others, nor does he openly endorse the alternative he does not reject. And indeed he seems to hint that quotations do have significant structure when he says, 'It is clear that we can correlate a structural-descriptive name with every quotation-mark name, one which is free from quotation marks and possesses the same extension (*i.e.* denotes the same expression) and vice versa' (157). It is difficult to see how the correlation could be established if we replaced each quotation by some other arbitrary symbol, as we could do if quotation-mark names were like the proper names of men.

Quine takes matters a bit further by asserting that although quotations are 'logically unstructured' (190), and of course expressions have non-referential occurrences inside quotation marks, still the latter feature is 'dispelled by an easy change in notation' (144) which leaves us with the logically structured devices of spelling and concatenation. The formula for the 'easy change' can apparently be given by a definition (189, 190). If this suggestion can be carried out, then the most recalcitrant aspect of quotation yields to theory, for the truth conditions for sentences containing quotations can be equated with the truth conditions for the sentences got from them by substituting for the quotations their definitional equivalents in the idiom of spelling. On such a theory, there is no longer an infinite number of semantical primitives, in spite of the fact that quotations cannot be shown to contain parts with independent semantical roles. If we accept a theory of this kind, we are forced to allow a species of structure that may not deserve to be called 'logical', but certainly is directly and indissolubly linked with the logical, a kind of structure missing in ordinary proper names.[15]

Second Example. Scheffler on Indirect Discourse. Israel Scheffler has proposed what he calls an inscriptional approach to indirect discourse.[16] Carnap at one time analysed sentences in indirect discourse as involving a relation (elaborated in terms of a notion of intensional isomorphism) between a speaker and a sentence.[17]

[15] Peter Geach has insisted, in *Mental Acts* and elsewhere, that a quotation 'must be taken as *describing* the expression in terms of its parts' (*Mental Acts*, 83). But he does not explain how a quotation can be assigned the structure of a description.

For a more extended discussion of quotation, see Essay 6.

[16] I. Scheffler, 'An Inscriptional Approach to Indirect Discourse' and *The Anatomy of Inquiry*.

[17] R. Carnap, *Meaning and Necessity*.

Church objected that Carnap's analysis carried an implicit reference to a language which was lacking in the sentences to be analysed, and added that he believed any correct analysis would interpret the that-clause as referring to a proposition.[18] Scheffler set out to show that a correct analysis could manage with less by demonstrating what could be done with an ontology of inscriptions (and utterances).

Scheffler suggests we analyse 'Tonkin said that snow is white' as 'Tonkin spoke a that-snow-is-white utterance'. Since an utterance or inscription belongs eternally to the language of its ephemeral producer, Church's reproach to Carnap has no force here. The essential part of the story, from the point of view of present concerns, is that the expression 'that-snow-is-white' is to be treated as a unitary predicate (of utterances or inscriptions). Scheffler calls the 'that' an operator which applies to a sentence to form a composite general term. 'Composite' cannot, in this use, mean 'logically complex'; not, at least, until more theory is forthcoming. As in the case of quotations, the syntax is clear enough (as it is in indirect discourse generally, give or take some rewriting of verbs and pronouns on principles easier to master than to describe). But there is no hint as to how the meaning of these predicates depends on their structure. Failing a theory, we must view each new predicate as a semantical primitive. Given their syntax, though (put any sentence after 'that' and spice with hyphens), it is obvious there are infinitely many such predicates, so languages with no more structure than Scheffler allows are, on my account, unlearnable. Even the claim that 'that' is a predicate-forming operator must be recognized as a purely syntactical comment that has no echo in the theory of meaning.

The possibility remains open, it may be countered, that a theory will yet be produced to reveal more structure. True. But if Scheffler wants the benefit of this possibility, he must pay for it be renouncing the claim to have shown 'the *logical form and ontological character*' of sentences in indirect discourse;[19] more theory may mean more ontology. If logical form tells us about ontology, then Quine's trick with quotations won't work, for the logical form of a quotation is that of a singular term without parts, and its manifest ontology is just the expression it names. But the definition that takes advantage

[18] A. Church, 'On Carnap's Analysis of Statements of Assertion and Belief'.
[19] I. Scheffler, *The Anatomy of Inquiry*, 101. His italics.

of its structure brings in more entities (those named in the spelling). It is only an accident that in this application the method draws on no entities not already named by some quotation.[20]

Third Example. Quine on Belief Sentences. We can always trade problems of ontology raised by putatively referential expressions and positions for problems of logical articulation. If an expression offend by its supposed reference, there is no need to pluck it out. It is enough to declare the expression a meaningless part of a meaningful expression. This treatment will not quite kill before it cures; after all singular terms and positions open to quantification have been welded into their contexts, there will remain the logical structure created by treating sentences as unanalysable units and by the pure sentential connectives. But semantics without ontology is not very interesting, and a language like our own for which no better could be done would be a paradigm of unlearnability.

Quine has not, of course, gone this far, but in an isolated moment he seems to come close. Near the close of a long, brilliant, and discouraging search for a satisfactory theory of belief sentences (and their relatives—sentences in indirect discourse, sentences about doubts, wonderings, fears, desires, and so forth) he remarks that once we give up trying to quantify over things believed

> . . . there is no need to recognize 'believes' and similar verbs as relative terms at all; no need to countenance their predicative use as in '*w* believes *x*' (as against '*w* believes that *p*'); no need, therefore, to see 'that *p*' as a term. Hence a final alternative that I find as appealing as any is simply to dispense with the objects of the propositional attitudes . . . This means viewing 'Tom believes [Cicero denounced Catiline]' . . . as of the form 'F*a*' with a = Tom and complex 'F'. The verb 'believes' here ceases to be a term and becomes part of an operator 'believes that' . . . which, applied to a sentence, produces a composite absolute general term whereof the sentence is counted an immediate constituent.[21]

In one respect, this goes a step beyond Scheffler, for even the main verb ('believes' in this case) is made inaccessible to logical analysis. Talk of constituents and operators must of course be taken as purely syntactical, without basis in semantical theory. If there is any element beyond syntax common to belief sentences as a class, Quine's account does not say what it is. And, of course, a language

[20] For further work on indirect discourse, see Essay 7.
[21] W. V. Quine, *Word and Object*, 215, 216.

for which no more theory can be given is, by my reckoning, unlearnable. Quine does not, however, rule out hope of further theory, perhaps something like what he offers for quotations. But in the case of belief sentences it is not easy to imagine the theory that could yield the required structure without adding to the ontology.

Fourth Example. Church on the Logic of Sense and Denotation. In the last two examples, loss of a desirable minimum of articulation of meaning seemed to be the result of overzealous attention to problems of ontology. But it would be a mistake to infer that by being prodigal with intensional entities we can solve all problems.

Frege asks us to suppose that certain verbs, like 'believes' (or 'believes that'), do double duty.[22] First, they create a context in which the words that follow come to refer to their usual sense or meaning. Second (assuming the verb is the main verb of the sentence), they perform a normal kind of duty by mapping persons and propositions on to truth values. This is a dark doctrine, particularly with respect to the first point, and Frege seems to have thought so himself. His view was, perhaps, that this is the best we can do for natural languages, but that in a logically more transparent language, different words would be used to refer to sense and to denotation, thus relieving the burdened verbs of their first, and more obscure, duty. But now it is only a few steps to an infinite primitive vocabulary. After the first appearances of verbs like 'believes' we introduce new expressions for senses. Next, we notice that there is no theory which interprets these new expressions as logically structured. A new vocabulary is again needed, along the same lines, each time we iterate 'believes'; and there is no limit to the number of possible iterations.

It should not be thought that there would be less trouble with Frege's original suggestion. Even supposing we made good sense of the idea that certain words create a context in which other words take on new meanings (an idea that only makes them sound like functors), there would remain the task of reducing to theory the determination of those meanings—an infinite number each for at least some words. The problem is not how the individual expressions that make up a sentence governed by 'believes', given the meanings they have in such a context, combine to denote a proposition; the

[22] See G. Frege, 'On Sense and Reference'.

problem is rather to state the rule that gives each the meaning it does have.

To return to our speculations concerning the form Frege might have given a language notationally superior to ordinary language, but like it in its capacity to deal with belief sentences and their kin: the features I attributed above to this language are to be found in a language proposed by Alonzo Church.[23] In Church's notation, the fact that the new expressions brought into play as we scale the semantic ladder are not logically complex is superficially obscured by their being syntactically composed from the expressions for the next level below, plus change of subscript. Things are so devised that 'if, in a well-formed formula without free variables, all the subscripts in all the type symbols appearing are increased by 1, the resulting well-formed formula denotes the sense of the first one' (17). But this rule cannot, of course, be exploited as part of a theory of meaning for the language; expression and subscript cannot be viewed as having independent meanings. Relative to the expression on any given level, the expressions for the level above are semantical primitives, as Church clearly indicates (8). I don't suggest any error on Church's part; he never hints that the case is other than I say. But I do submit that Church's language of Sense and Denotation is, even in principle, unlearnable.

[23] A. Church, 'A Formulation of the Logic of Sense and Denotation'.

2 *Truth and Meaning*

It is conceded by most philosophers of language, and recently by some linguists, that a satisfactory theory of meaning must give an account of how the meanings of sentences depend upon the meanings of words. Unless such an account could be supplied for a particular language, it is argued, there would be no explaining the fact that we can learn the language: no explaining the fact that, on mastering a finite vocabulary and a finitely stated set of rules, we are prepared to produce and to understand any of a potential infinitude of sentences. I do not dispute these vague claims, in which I sense more than a kernal of truth.[1] Instead I want to ask what it is for a theory to give an account of the kind adumbrated.

One proposal is to begin by assigning some entity as meaning to each word (or other significant syntactical feature) of the sentence; thus we might assign Theaetetus to 'Theaetetus' and the property of flying to 'flies' in the sentence 'Theaetetus flies'. The problem then arises how the meaning of the sentence is generated from these meanings. Viewing concatenation as a significant piece of syntax, we may assign to it the relation of participating in or instantiating; however, it is obvious that we have here the start of an infinite regress. Frege sought to avoid the regress by saying that the entities corresponding to predicates (for example) are 'unsaturated' or 'incomplete' in contrast to the entities that correspond to names, but this doctrine seems to label a difficulty rather than solve it.

The point will emerge if we think for a moment of complex singular terms, to which Frege's theory applies along with sentences. Consider the expression 'the father of Annette'; how does the

[1] See Essay 1.

meaning of the whole depend on the meaning of the parts? The answer would seem to be that the meaning of 'the father of' is such that when this expression is prefixed to a singular term the result refers to the father of the person to whom the singular term refers. What part is played, in this account, by the unsaturated or incomplete entity for which 'the father of' stands? All we can think to say is that this entity 'yields' or 'gives' the father of x as value when the argument is x, or perhaps that this entity maps people on to their fathers. It may not be clear whether the entity for which 'the father of' is said to stand performs any genuine explanatory function as long as we stick to individual expressions; so think instead of the infinite class of expressions formed by writing 'the father of' zero or more times in front of 'Annette'. It is easy to supply a theory that tells, for an arbitrary one of these singular terms, what it refers to: if the term is 'Annette' it refers to Annette, while if the term is complex, consisting of 'the father of' prefixed to a singular term t, then it refers to the father of the person to whom t refers. It is obvious that no entity corresponding to 'the father of' is, or needs to be, mentioned in stating this theory.

It would be inappropriate to complain that this little theory *uses* the words 'the father of' in giving the reference of expressions containing those words. For the task was to give the meaning of all expressions in a certain infinite set on the basis of the meaning of the parts; it was not in the bargain also to give the meanings of the atomic parts. On the other hand, it is now evident that a satisfactory theory of the meanings of complex expressions may not require entities as meanings of all the parts. It behoves us then to rephrase our demand on a satisfactory theory of meaning so as not to suggest that individual words must have meanings at all, in any sense that transcends the fact that they have a systematic effect on the meanings of the sentences in which they occur. Actually, for the case at hand we can do better still in stating the criterion of success: what we wanted, and what we got, is a theory that entails every sentence of the form 't refers to x' where 't' is replaced by a structural description[2] of a singular term, and 'x' is replaced by that term itself. Further, our theory accomplishes this without appeal to any semantical concepts beyond the basic 'refers to'. Finally, the theory

[2] A 'structural description' of an expression describes the expression as a concatenation of elements drawn from a fixed finite list (for example of words or letters).

clearly suggests an effective procedure for determining, for any singular term in its universe, what that term refers to.

A theory with such evident merits deserves wider application. The device proposed by Frege to this end has a brilliant simplicity: count predicates as a special case of functional expressions, and sentences as a special case of complex singular terms. Now, however, a difficulty looms if we want to continue in our present (implicit) course of identifying the meaning of a singular term with its reference. The difficulty follows upon making two reasonable assumptions: that logically equivalent singular terms have the same reference, and that a singular term does not change its reference if a contained singular term is replaced by another with the same reference. But now suppose that 'R' and 'S' abbreviate any two sentences alike in truth value. Then the following four sentences have the same reference:

(1) R
(2) $\hat{x}(x = x . R) = \hat{x}(x = x)$
(3) $\hat{x}(x = x . S) = \hat{x}(x = x)$
(4) S

For (1) and (2) are logically equivalent, as are (3) and (4), while (3) differs from (2) only in containing the singular term '$\hat{x}(x = x . S)$' where (2) contains '$\hat{x}(x = x . R)$' and these refer to the same thing if S and R are alike in truth value. Hence any two sentences have the same reference if they have the same truth value.[3] And if the meaning of a sentence is what it refers to, all sentences alike in truth value must be synonymous—an intolerable result.

Apparently we must abandon the present approach as leading to a theory of meaning. This is the natural point at which to turn for help to the distinction between meaning and reference. The trouble, we are told, is that questions of reference are, in general, settled by extra-linguistic facts, questions of meaning not, and the facts can conflate the references of expressions that are not synonymous. If we want a theory that gives the meaning (as distinct from reference) of each sentence, we must start with the meaning (as distinct from reference) of the parts.

Up to here we have been following in Frege's footsteps; thanks to

[3] The argument derives from Frege. See A. Church, *Introduction to Mathematical Logic*, 24–5. It is perhaps worth mentioning that the argument does not depend on any particular identification of the entities to which sentences are supposed to refer.

him, the path is well known and even well worn. But now, I would like to suggest, we have reached an impasse: the switch from reference to meaning leads to no useful account of how the meanings of sentences depend upon the meanings of the words (or other structural features) that compose them. Ask, for example, for the meaning of 'Theaetetus flies'. A Fregean answer might go something like this: given the meaning of 'Theaetetus' as argument, the meaning of 'flies' yields the meaning of 'Theaetetus flies' as value. The vacuity of this answer is obvious. We wanted to know what the meaning of 'Theaetetus flies' is; it is no progress to be told that it is the meaning of 'Theaetetus flies'. This much we knew before any theory was in sight. In the bogus account just given, talk of the structure of the sentence and of the meanings of words was idle, for it played no role in producing the given description of the meaning of the sentence.

The contrast here between a real and pretended account will be plainer still if we ask for a theory, analogous to the miniature theory of reference of singular terms just sketched, but different in dealing with meanings in place of references. What analogy demands is a theory that has as consequences all sentences of the form '*s* means *m*' where '*s*' is replaced by a structural description of a sentence and '*m*' is replaced by a singular term that refers to the meaning of that sentence; a theory, moreover, that provides an effective method for arriving at the meaning of an arbitrary sentence structurally described. Clearly some more articulate way of referring to meanings that any we have seen is essential if these criteria are to be met.[4] Meanings as entities, or the related concept of synonymy, allow us to formulate the following rule relating sentences and their parts: sentences are synonymous whose corresponding parts are synonymous ('corresponding' here needs spelling out of course). And meanings as entities may, in theories such as Frege's, do duty, on occasion, as references, thus losing their status as entities distinct from references. Paradoxically, the one thing meanings do not seem to do is oil the wheels of a theory of meaning—at least as long as we require of such a theory that it non-trivially give the meaning of

[4] It may be thought that Church, in 'A Formulation of the Logic of Sense and Denotation', has given a theory of meaning that makes essential use of meanings as entities. But this is not the case: Church's logics of sense and denotation are interpreted as being about meanings, but they do not mention expressions and so cannot of course be theories of meaning in the sense now under discussion.

every sentence in the language. My objection to meanings in the theory of meaning is not that they are abstract or that their identity conditions are obscure, but that they have no demonstrated use.

This is the place to scotch another hopeful thought. Suppose we have a satisfactory theory of syntax for our language, consisting of an effective method of telling, for an arbitrary expression, whether or not it is independently meaningful (i.e. a sentence), and assume as usual that this involves viewing each sentence as composed, in allowable ways, out of elements drawn from a fixed finite stock of atomic syntactical elements (roughly, words). The hopeful thought is that syntax, so conceived, will yield semantics when a dictionary giving the meaning of each syntactic atom is added. Hopes will be dashed, however, if semantics is to comprise a theory of meaning in our sense, for knowledge of the structural characteristics that make for meaningfulness in a sentence, plus knowledge of the meanings of the ultimate parts, does not add up to knowledge of what a sentence means. The point is easily illustrated by belief sentences. Their syntax is relatively unproblematic. Yet, adding a dictionary does not touch the standard semantic problem, which is that we cannot account for even as much as the truth conditions of such sentences on the basis of what we know of the meanings of the words in them. The situation is not radically altered by refining the dictionary to indicate which meaning or meanings an ambiguous expression bears in each of its possible contexts; the problem of belief sentences persists after ambiguities are resolved.

The fact that recursive syntax with dictionary added is not necessarily recursive semantics has been obscured in some recent writing on linguistics by the intrusion of semantic criteria into the discussion of purportedly syntactic theories. The matter would boil down to a harmless difference over terminology if the semantic criteria were clear; but they are not. While there is agreement that it is the central task of semantics to give the semantic interpretation (the meaning) of every sentence in the language, nowhere in the linguistic literature will one find, so far as I know, a straightforward account of how a theory performs this task, or how to tell when it has been accomplished. The contrast with syntax is striking. The main job of a modest syntax is to characterize *meaningfulness* (or sentencehood). We may have as much confidence in the correctness of such a characterization as we have in the representativeness of our sample and our ability to say when particular expressions are meaningful (sentences). What clear and analogous task and test exist for semantics?[5]

We decided a while back not to assume that parts of sentences have meanings except in the ontologically neutral sense of making a systematic contribution to the meaning of the sentences in which they occur. Since postulating meanings has netted nothing, let us return to that insight. One direction in which it points is a certain holistic view of meaning. If sentences depend for their meaning on their structure, and we understand the meaning of each item in the structure only as an abstraction from the totality of sentences in which it features, then we can give the meaning of any sentence (or word) only by giving the meaning of every sentence (and word) in the language. Frege said that only in the context of a sentence does a word have meaning; in the same vein he might have added that only in the context of the language does a sentence (and therefore a word) have meaning.

This degree of holism was already implicit in the suggestion that an adequate theory of meaning must entail *all* sentences of the form '*s* means *m*'. But now, having found no more help in meanings of sentences than in meanings of words, let us ask whether we can get rid of the troublesome singular terms supposed to replace '*m*' and to refer to meanings. In a way, nothing could be easier: just write '*s* means that *p*', and imagine '*p*' replaced by a sentence. Sentences, as we have seen, cannot name meanings, and sentences with 'that' prefixed are not names at all, unless we decide so. It looks as though we are in trouble on another count, however, for it is reasonable to expect that in wrestling with the logic of the apparently non-extensional 'means that' we will encounter problems as hard as, or perhaps identical with, the problems our theory is out to solve.

The only way I know to deal with this difficulty is simple, and radical. Anxiety that we are enmeshed in the intensional springs from using the words 'means that' as filling between description of

[5] For a recent statement of the role of semantics in linguistics, see Noam Chomsky, 'Topics in the Theory of Generative Grammar'. In this article, Chomsky (1) emphasizes the central importance of semantics in linguistic theory, (2) argues for the superiority of transformational grammars over phrase-structure grammars largely on the grounds that, although phrase-structure grammars may be adequate to define sentencehood for (at least) some natural languages, they are inadequate as a foundation for semantics, and (3) comments repeatedly on the 'rather primitive state' of the concepts of semantics and remarks that the notion of semantic interpretation 'still resists any deep analysis'.

sentence and sentence, but it may be that the success of our venture depends not on the filling but on what it fills. The theory will have done its work if it provides, for every sentence *s* in the language under study, a matching sentence (to replace '*p*') that, in some way yet to be made clear, 'gives the meaning' of *s*. One obvious candidate for matching sentence is just *s* itself, if the object language is contained in the metalanguage; otherwise a translation of *s* in the metalanguage. As a final bold step, let us try treating the position occupied by '*p*' extensionally: to implement this, sweep away the obscure 'means that', provide the sentence that replaces '*p*' with a proper sentential connective, and supply the description that replaces '*s*' with its own predicate. The plausible result is

(*T*) *s* is *T* if and only if *p*.

What we require of a theory of meaning for a language *L* is that without appeal to any (further) semantical notions it place enough restrictions on the predicate 'is *T*' to entail all sentences got from schema *T* when '*s*' is replaced by a structural description of a sentence of *L* and '*p*' by that sentence.

Any two predicates satisfying this condition have the same extension,[6] so if the metalanguage is rich enough, nothing stands in the way of putting what I am calling a theory of meaning into the form of an explicit definition of a predicate 'is *T*'. But whether explicitly defined or recursively characterized, it is clear that the sentences to which the predicate 'is *T*' applies will be just the true sentences of *L*, for the condition we have placed on satisfactory theories of meaning is in essence Tarski's Convention *T* that tests the adequacy of a formal semantical definition of truth.[7]

The path to this point has been tortuous, but the conclusion may be stated simply: a theory of meaning for a language *L* shows 'how the meanings of sentences depend upon the meanings of words' if it contains a (recursive) definition of truth-in-L. And, so far at least, we have no other idea how to turn the trick. It is worth emphasizing that the concept of truth played no ostensible role in stating our original problem. That problem, upon refinement, led to the view that an adequate theory of meaning must characterize a predicate meeting certain conditions. It was in the nature of a discovery that

[6] Assuming, of course, that the extension of these predicates is limited to the sentences of *L*.

[7] A. Tarski, 'The Concept of Truth in Formalized Languages'.

such a predicate would apply exactly to the true sentences. I hope that what I am saying may be described in part as defending the philosophical importance of Tarski's semantical concept of truth. But my defence is only distantly related, if at all, to the question whether the concept Tarski has shown how to define is the (or a) philosophically interesting conception of truth, or the question whether Tarski has cast any light on the ordinary use of such words as 'true' and 'truth'. It is a misfortune that dust from futile and confused battles over these questions has prevented those with a theoretical interest in language—philosophers, logicians, psychologists, and linguists alike—from seeing in the semantical concept of truth (under whatever name) the sophisticated and powerful foundation of a competent theory of meaning.

There is no need to suppress, of course, the obvious connection between a definition of truth of the kind Tarski has shown how to construct, and the concept of meaning. It is this: the definition works by giving necessary and sufficient conditions for the truth of every sentence, and to give truth conditions is a way of giving the meaning of a sentence. To know the semantic concept of truth for a language is to know what it is for a sentence—any sentence—to be true, and this amounts, in one good sense we can give to the phrase, to understanding the language. This at any rate is my excuse for a feature of the present discussion that is apt to shock old hands; my freewheeling use of the word 'meaning', for what I call a theory of meaning has after all turned out to make no use of meanings, whether of sentences or of words. Indeed, since a Tarski-type truth definition supplies all we have asked so far of a theory of meaning, it is clear that such a theory falls comfortably within what Quine terms the 'theory of reference' as distinguished from what he terms the 'theory of meaning'. So much to the good for what I call a theory of meaning, and so much, perhaps, against my so calling it.[8]

A theory of meaning (in my mildly perverse sense) is an empirical theory, and its ambition is to account for the workings of a natural language. Like any theory, it may be tested by comparing some of its consequences with the facts. In the present case this is easy, for the

[8] But Quine may be quoted in support of my usage: '. . . in point of *meaning* . . . a word may be said to be determined to whatever extent the truth or falsehood of its contexts is determined.' ('Truth by Convention', 82.) Since a truth definition determines the truth value of every sentence in the object language (relative to a sentence in the metalanguage), it determines the meaning of every word and sentence. This would seem to justify the title Theory of Meaning.

theory has been characterized as issuing in an infinite flood of sentences each giving the truth conditions of a sentence; we only need to ask, in sample cases, whether what the theory avers to be the truth conditions for a sentence really are. A typical test case might involve deciding whether the sentence 'Snow is white' *is* true if and only if snow is white. Not all cases will be so simple (for reasons to be sketched), but it is evident that this sort of test does not invite counting noses. A sharp conception of what constitutes a theory in this domain furnishes an exciting context for raising deep questions about when a theory of language is correct and how it is to be tried. But the difficulties are theoretical, not practical. In application, the trouble is to get a theory that comes close to working; anyone can tell whether it is right.[9] One can see why this is so. The theory reveals nothing new about the conditions under which an individual sentence is true; it does not make those conditions any clearer than the sentence itself does. The work of the theory is in relating the known truth conditions of each sentence to those aspects ('words') of the sentence that recur in other sentences, and can be assigned identical roles in other sentences. Empirical power in such a theory depends on success in recovering the structure of a very complicated ability—the ability to speak and understand a language. We can tell easily enough when particular pronouncements of the theory comport with our understanding of the language; this is consistent with a feeble insight into the design of the machinery of our linguistic accomplishments.

The remarks of the last paragraph apply directly only to the special case where it is assumed that the language for which truth is being characterized is part of the language used and understood by the characterizer. Under these circumstances, the framer of a theory will as a matter of course avail himself when he can of the built-in convenience of a metalanguage with a sentence guaranteed equivalent to each sentence in the object language. Still, this fact ought not to con us into thinking a theory any more correct that entails '"Snow is white" is true if and only if snow is white' than one that entails instead:

(*S*) 'Snow is white' is true if and only if grass is green,

[9] To give a single example: it is clearly a count in favour of a theory that it entails '"Snow is white" is true if and only if snow is white'. But to contrive a theory that entails this (and works for all related sentences) is not trivial. I do not know a wholly satisfactory theory that succeeds with this very case (the problem of 'mass terms').

provided, of course, we are as sure of the truth of (*S*) as we are of that of its more celebrated predecessor. Yet (*S*) may not encourage the same confidence that a theory that entails it deserves to be called a theory of meaning.

The threatened failure of nerve may be counteracted as follows. The grotesqueness of (*S*) is in itself nothing against a theory of which it is a consequence, provided the theory gives the correct results for every sentence (on the basis of its structure, there being no other way). It is not easy to see how (*S*) could be party to such an enterprise, but if it were—if, that is, (*S*) followed from a characterization of the predicate 'is true' that led to the invariable pairing of truths with truths and falsehoods with falsehoods—then there would not, I think, be anything essential to the idea of meaning that remained to be captured.[10]

What appears to the right of the biconditional in sentences of the form '*s* is true if and only if *p*' when such sentences are consequences of a theory of truth plays its role in determining the meaning of *s* not by pretending synonymy but by adding one more brush-stroke to the picture which, taken as a whole, tells what there is to know of the meaning of *s*; this stroke is added by virtue of the fact that the sentence that replaces '*p*' is true if and only if *s* is.

It may help to reflect that (*S*) is acceptable, if it is, because we are independently sure of the truth of 'Snow is white' and 'Grass is green'; but in cases where we are unsure of the truth of a sentence, we can have confidence in a characterization of the truth predicate only if it pairs that sentence with one we have good reason to believe equivalent. It would be ill advised for someone who had any doubts about the colour of snow or grass to accept a theory that yielded (*S*), even if his doubts were of equal degree, unless he thought the colour of the one was tied to the colour of the other.[11] Omniscience can

[10] Critics have often failed to notice the essential proviso mentioned in this paragraph. The point is that (*S*) could not belong to any reasonably simple theory that also gave the right truth conditions for 'That is snow' and 'This is white'. (See the discussion of indexical expressions below.) [Footnote added in 1982.]

[11] This paragraph is confused. What it should say is that sentences of the theory are empirical generalizations about speakers, and so must not only be true but also lawlike. (*S*) presumably is not a law, since it does not support appropriate counterfactuals. It's also important that the evidence for accepting the (time and speaker relativized) truth conditions for 'That is snow' is based on the causal connection between a speaker's assent to the sentence and the demonstrative presentation of snow. For further discussion see Essay 12. [Footnote added in 1982.]

obviously afford more bizzare theories of meaning than ignorance; but then, omniscience has less need of communication.

It must be possible, of course, for the speaker of one language to construct a theory of meaning for the speaker of another, though in this case the empirical test of the correctness of the theory will no longer be trivial. As before, the aim of theory will be an infinite correlation of sentences alike in truth. But this time the theory-builder must not be assumed to have direct insight into likely equivalences between his own tongue and the alien. What he must do is find out, however he can, what sentences the alien holds true in his own tongue (or better, to what degree he holds them true). The linguist then will attempt to construct a characterization of truth-for-the-alien which yields, so far as possible, a mapping of sentences held true (or false) by the alien on to sentences held true (or false) by the linguist. Supposing no perfect fit is found, the residue of sentences held true translated by sentences held false (and vice versa) is the margin for error (foreign or domestic). Charity in interpreting the words and thoughts of others is unavoidable in another direction as well: just as we must maximize agreement, or risk not making sense of what the alien is talking about, so we must maximize the self-consistency we attribute to him, on pain of not understanding *him*. No single principle of optimum charity emerges; the constraints therefore determine no single theory. In a theory of radical translation (as Quine calls it) there is no completely disentangling questions of what the alien means from questions of what he believes. We do not know what someone means unless we know what he believes; we do not know what someone believes unless we know what he means. In radical interpretation we are able to break into this circle, if only incompletely, because we can sometimes tell that a person accedes to a sentence we do not understand.[12]

In the past few pages I have been asking how a theory of meaning that takes the form of a truth definition can be empirically tested, and have blithely ignored the prior question whether there is any serious chance such a theory can be given for a natural language. What are the prospects for a formal semantical theory of a natural

[12] This sketch of how a theory of meaning for an alien tongue can be tested obviously owes it inspiration to Quine's account of radical translation in Chapter II of *Word and Object*. In suggesting that an acceptable theory of radical translation take the form of a recursive characterization of truth, I go beyond Quine. Toward the end of this paper, in the discussion of demonstratives, another strong point of agreement will turn up.

language? Very poor, according to Tarski; and I believe most logicians, philosophers of language, and linguists agree.[13] Let me do what I can to dispel the pessimism. What I can in a general and programmatic way, of course, for here the proof of the pudding will certainly be in the proof of the right theorems.

Tarski concludes the first section of his classic essay on the concept of truth in formalized languages with the following remarks, which he italicizes:

> ... *The very possibility of a consistent use of the expression 'true sentence' which is in harmony with the laws of logic and the spirit of everyday language seems to be very questionable, and consequently the same doubt attaches to the possibility of constructing a correct definition of this expression.* (165)

Late in the same essay, he returns to the subject:

> ... the concept of truth (as well as other semantical concepts) when applied to colloquial language in conjunction with the normal laws of logic leads inevitably to confusions and contradictions. Whoever wishes, in spite of all difficulties, to pursue the semantics of colloquial language with the help of exact methods will be driven first to undertake the thankless task of a reform of this language. He will find it necessary to define its structure, to overcome the ambiguity of the terms which occur in it, and finally to split the language into a series of languages of greater and greater extent, each of which stands in the same relation to the next in which a formalized language stands to its metalanguage. It may, however be doubted whether the language of everyday life, after being 'rationalized' in this way, would still preserve its naturalness and whether it would not rather take on the characteristic features of the formalized languages. (267)

Two themes emerge: that the universal character of natural languages leads to contradiction (the semantic paradoxes), and that natural languages are too confused and amorphous to permit the direct application of formal methods. The first point deserves a serious answer, and I wish I had one. As it is, I will say only why I think we are justified in carrying on without having disinfected this particular source of conceptual anxiety. The semantic paradoxes arise when the range of the quantifiers in the object language is too generous in certain ways. But it is not really clear how unfair to Urdu or to Wendish it would be to view the range of their quantifiers

[13] So far as I am aware, there has been very little discussion of whether a formal truth definition can be given for a natural language. But in a more general vein, several people have urged that the concepts of formal semantics be applied to natural language. See, for example, the contributions of Yehoshua Bar-Hillel and Evert Beth to *The Philosophy of Rudolph Carnap*, and Bar-Hillel's 'Logical Syntax and Semantics'.

as insufficient to yield an explicit definition of 'true-in-Urdu' or 'true-in-Wendish'. Or, to put the matter in another, if not more serious way, there may in the nature of the case always be something we grasp in understanding the language of another (the concept of truth) that we cannot communicate to him. In any case, most of the problems of general philosophical interest arise within a fragment of the relevant natural language that may be conceived as containing very little set theory. Of course these comments do not meet the claim that natural languages are universal. But it seems to me that this claim, now that we know such universality leads to paradox, is suspect.

Tarski's second point is that we would have to reform a natural language out of all recognition before we could apply formal semantical methods. If this is true, it is fatal to my project, for the task of a theory of meaning as I conceive it is not to change, improve, or reform a language, but to describe and understand it. Let us look at the positive side. Tarski has shown the way to giving a theory for interpreted formal languages of various kinds; pick one as much like English as possible. Since this new language has been explained in English and contains much English we not only may, but I think must, view it as part of English for those who understand it. For this fragment of English we have, *ex hypothesi*, a theory of the required sort. Not only that, but in interpreting this adjunct of English in old English we necessarily gave hints connecting old and new. Wherever there are sentences of old English with the same truth conditions as sentences in the adjunct we may extend the theory to cover them. Much of what is called for is to mechanize as far as possible what we now do by art when we put ordinary English into one or another canonical notation. The point is not that canonical notation is better than the rough original idiom, but rather that if we know what idiom the canonical notation is canonical *for*, we have as good a theory for the idiom as for its kept companion.

Philosophers have long been at the hard work of applying theory to ordinary language by the device of matching sentences in the vernacular with sentences for which they have a theory. Frege's massive contribution was to show how 'all', 'some', 'every', 'each', 'none', and associated pronouns, in some of their uses, could be tamed; for the first time, it was possible to dream of a formal semantics for a significant part of a natural language. This dream came true in a sharp way with the work of Tarski. It would be a shame to miss the fact that as a result of these two magnificent achievements, Frege's and Tarski's, we have gained a deep insight into the structure of our mother tongues. Philosophers of a logical bent have tended to start where the theory was and work out towards the complications of natural language. Contemporary linguists, with an aim that cannot easily be seen to be different, start with the ordinary and work toward a general theory. If either party is successful, there must be a meeting. Recent work by Chomsky and others is doing much to bring the complexities of natural languages within the scope of serious theory. To give an example: suppose success in giving the truth conditions for some significant range of sentences in the active voice. Then with a formal procedure for transforming each such sentence into a corresponding sentence in the passive voice, the theory of truth could be extended in an obvious way to this new set of sentences.[14]

One problem touched on in passing by Tarski does not, at least in all its manifestations, have to be solved to get ahead with theory: the existence in natural languages of 'ambiguous terms'. As long as ambiguity does not affect grammatical form, and can be translated, ambiguity for ambiguity, into the metalanguage, a truth definition will not tell us any lies. The chief trouble, for systematic semantics, with the phrase 'believes that' in English lies not in its vagueness, ambiguity, or unsuitability for incorporation in a serious science: let our metalanguage be English, and all *these* problems will be carried without loss or gain into the metalanguage. But the central problem of the logical grammar of 'believes that' will remain to haunt us.

The example is suited to illustrating another, and related, point, for the discussion of belief sentences has been plagued by failure to

[14] The *rapprochement* I prospectively imagine between transformational grammar and a sound theory of meaning has been much advanced by a recent change in the conception of transformational grammar described by Chomsky in the article referred to above (note 5). The structures generated by the phrase-structure part of the grammar, it has been realized for some time, are those suited to semantic interpretation; but this view is inconsistent with the idea, held by Chomsky until recently, that recursive operations are introduced only by the transformation rules. Chomsky now believes the phrase-structure rules are recursive. Since languages to which formal semantic methods directly and naturally apply are ones for which a (recursive) phrase-structure grammar is appropriate, it is clear that Chomsky's present picture of the relation between the structures generated by the phrase-structure part of the grammar, and the sentences of the language, is very much like the picture many logicians and philosophers have had of the relation between the richer formalized languages and ordinary language. (In these remarks I am indebted to Bruce Vermazen.)

observe a fundamental distinction between tasks: uncovering the logical grammar or form of sentences (which is in the province of a theory of meaning as I construe it), and the analysis of individual words or expressions (which are treated as primitive by the theory). Thus Carnap, in the first edition of *Meaning and Necessity*, suggested we render 'John believes that the earth is round' as 'John responds affirmatively to "the earth is round" as an English sentence'. He gave this up when Mates pointed out that John might respond affirmatively to one sentence and not to another no matter how close in meaning.[15] But there is a confusion here from the start. The semantic structure of a belief sentence, according to this idea of Carnap's, is given by a three-place predicate with places reserved for expressions referring to a person, a sentence, and a language. It is a different sort of problem entirely to attempt an analysis of this predicate, perhaps along behaviouristic lines. Not least among the merits of Tarski's conception of a theory of truth is that the purity of method it demands of us follows from the formulation of the problem itself, not from the self-imposed restraint of some adventitious philosophical puritanism.

I think it is hard to exaggerate the advantages to philosophy of language of bearing in mind this distinction between questions of logical form or grammar, and the analysis of individual concepts. Another example may help advertise the point.

If we suppose questions of logical grammar settled, sentences like 'Bardot is good' raise no special problems for a truth definition. The deep differences between descriptive and evaluative (emotive, expressive, etc.) terms do not show here. Even if we hold there is some important sense in which moral or evaluative sentences do not have a truth value (for example, because they cannot be verified), we ought not to boggle at '"Bardot is good" is true if and only if Bardot is good'; in a theory of truth, this consequence should follow with the rest, keeping track, as must be done, of the semantic location of such sentences in the language as a whole—of their relation to generalizations, their role in such compound sentences as 'Bardot is good and Bardot is foolish', and so on. What is special to evaluative words is simply not touched: the mystery is transferred from the word 'good' in the object language to its translation in the metalanguage.

[15] B. Mates, 'Synonymity

But 'good' as it features in 'Bardot is a good actress' is another matter. The problem is not that the translation of this sentence is not in the metalanguage—let us suppose it is. The problem is to frame a truth definition such that '"Bardot is a good actress" is true if and only if Bardot is a good actress'—and all other sentences like it—are consequences. Obviously 'good actress' does not mean 'good and an actress'. We might think of taking 'is a good actress' as an unanalysed predicate. This would obliterate all connection between 'is a good actress' and 'is a good mother', and it would give us no excuse to think of 'good', in these uses, as a word or semantic element. But worse, it would bar us from framing a truth definition at all, for there is no end to the predicates we would have to treat as logically simple (and hence accommodate in separate clauses in the definition of satisfaction): 'is a good companion to dogs', 'is a good 28-years old conversationalist', and so forth. The problem is not peculiar to the case: it is the problem of attributive adjectives generally.

It is consistent with the attitude taken here to deem it usually a strategic error to undertake philosophical analysis of words or expressions which is not preceded by or at any rate accompanied by the attempt to get the logical grammar straight. For how can we have any confidence in our analyses of words like 'right', 'ought', 'can', and 'obliged', or the phrases we use to talk of actions, events, and causes, when we do not know what (logical, semantical) parts of speech we have to deal with? I would say much the same about studies of the 'logic' of these and other words, and the sentences containing them. Whether the effort and ingenuity that have gone into the study of deontic logics, modal logics, imperative and erotetic logics have been largely futile or not cannot be known until we have acceptable semantic analyses of the sentences such systems purport to treat. Philosophers and logicians sometimes talk or work as if they were free to choose between, say, the truth-functional conditional and others, or free to introduce non-truth-functional sentential operators like 'Let it be the case that' or 'It ought to be the case that'. But in fact the decision is crucial. When we depart from idioms we can accommodate in a truth definition, we lapse into (or create) language for which we have no coherent semantical account—that is, no account at all of how such talk can be integrated into the language as a whole.

To return to our main theme: we have recognized that a theory of the kind proposed leaves the whole matter of what individual words mean exactly where it was. Even when the metalanguage is different from the object language, the theory exerts no pressure for improvement, clarification, or analysis of individual words, except when, by accident of vocabulary, straightforward translation fails. Just as synonomy, as between expressions, goes generally untreated, so also synonomy of sentences, and analyticity. Even such sentences as 'A vixen is a female fox' bear no special tag unless it is our pleasure to provide it. A truth definition does not distinguish between analytic sentences and others, except for sentences that owe their truth to the presence alone of the constants that give the theory its grip on structure: the theory entails not only that these sentences are true but that they will remain true under all significant rewritings of their non-logical parts. A notion of logical truth thus given limited application, related notions of logical equivalence and entailment will tag along. It is hard to imagine how a theory of meaning could fail to read a logic into its object language to this degree; and to the extent that it does, our intuitions of logical truth, equivalence, and entailment may be called upon in constructing and testing the theory.

I turn now to one more, and very large, fly in the ointment: the fact that the same sentence may at one time or in one mouth be true and at another time or in another mouth be false. Both logicians and those critical of formal methods here seem largely (though by no means universally) agreed that formal semantics and logic are incompetent to deal with the disturbances caused by demonstratives. Logicians have often reacted by downgrading natural language and trying to show how to get along without demonstratives; their critics react by downgrading logic and formal semantics. None of this can make me happy: clearly demonstratives cannot be eliminated from a natural language without loss or radical change, so there is no choice but to accommodate theory to them.

No logical errors result if we simply treat demonstratives as constants;[16] neither do any problems arise for giving a semantic truth definition. '"I am wise" is true if and only if I am wise', with its bland ignoring of the demonstrative element in 'I' comes off the assembly line along with '"Socrates is wise" is true if and only if Socrates is wise' with *its* bland indifference to the demonstrative element in 'is wise' (the tense).

[16] See W. V. Quine, *Methods of Logic*, 8.

What suffers in this treatment of demonstratives is not the definition of a truth predicate, but the plausibility of the claim that what has been defined is truth. For this claim is acceptable only if the speaker and circumstances of utterance of each sentence mentioned in the definition is matched by the speaker and circumstances of utterance of the truth definition itself. It could also be fairly pointed out that part of understanding demonstratives is knowing the rules by which they adjust their reference to circumstance; assimilating demonstratives to constant terms obliterates this feature. These complaints can be met, I think, though only by a fairly far-reaching revision in the theory of truth. I shall barely suggest how this could be done, but bare suggestion is all that is needed: the idea is technically trivial, and in line with work being done on the logic of the tenses.[17]

We could take truth to be a property, not of sentences, but of utterances, or speech acts, or ordered triples of sentences, times, and persons; but it is simplest just to view truth as a relation between a sentence, a person, and a time. Under such treatment, ordinary logic as now read applies as usual, but only to sets of sentences relativized to the same speaker and time; further logical relations between sentences spoken at different times and by different speakers may be articulated by new axioms. Such is not my concern. The theory of meaning undergoes a systematic but not puzzling change; corresponding to each expression with a demonstrative element there must in the theory be a phrase that relates the truth conditions of sentences in which the expression occurs to changing times and speakers. Thus the theory will entail sentences like the following:

> 'I am tired' is true as (potentially) spoken by p at t if and only if p is tired at t.
>
> 'That book was stolen' is true as (potentially) spoken by p at t if and only if the book demonstrated by p at t is stolen prior to t.[18]

Plainly, this course does not show how to eliminate demonstratives; for example, there is no suggestion that 'the book demonstrated by the speaker' can be substituted ubiquitously for 'that book' *salva veritate*. The fact that demonstratives are amenable to

[17] This claim has turned out to be naïvely optimistic. For some serious work on the subject, see S. Weinstein, 'Truth and Demonstratives'. [Note added in 1982.]

[18] There is more than an intimation of this approach to demonstratives and truth in J. L. Austin, 'Truth'.

formal treatment ought greatly to improve hopes for a serious semantics of natural language, for it is likely that many outstanding puzzles, such as the analysis of quotations or sentences about propositional attitudes, can be solved if we recognize a concealed demonstrative construction.

Now that we have relativized truth to times and speakers, it is appropriate to glance back at the problem of empirically testing a theory of meaning for an alien tongue. The essence of the method was, it will be remembered, to correlate held-true sentences with held-true sentences by way of a truth definition, and within the bounds of intelligible error. Now the picture must be elaborated to allow for the fact that sentences are true, and held true, only relative to a speaker and a time. Sentences with demonstratives obviously yield a very sensitive test of the correctness of a theory of meaning, and constitute the most direct link between language and the recurrent macroscopic objects of human interest and attention.[19]

In this paper I have assumed that the speakers of a language can effectively determine the meaning or meanings of an arbitrary expression (if it has a meaning), and that it is the central task of a theory of meaning to show how this is possible. I have argued that a characterization of a truth predicate describes the required kind of structure, and provides a clear and testable criterion of an adequate semantics for a natural language. No doubt there are other reasonable demands that may be put on a theory of meaning. But a theory that does no more than define truth for a language comes far closer to constituting a complete theory of meaning than superficial analysis might suggest; so, at least, I have urged.

Since I think there is no alternative, I have taken an optimistic and programmatic view of the possibilities for a formal characterization of a truth predicate for a natural language. But it must be allowed that a staggering list of difficulties and conundrums remains. To name a few: we do not know the logical form of counterfactual or subjunctive sentences; nor of sentences about probabilities and about causal relations; we have no good idea what the logical role of adverbs is, nor the role of attributive adjectives; we have no theory for mass terms like 'fire', 'water', and 'snow', nor for sentences about

[19] These remarks derive from Quine's idea that 'occasion sentences' (those with a demonstrative element) must play a central role in constructing a translation manual.

belief, perception, and intention, nor for verbs of action that imply purpose. And finally, there are all the sentences that seem not to have truth values at all: the imperatives, optatives, interrogatives, and a host more. A comprehensive theory of meaning for a natural language must cope successfully with each of these problems.[20]

[20] For attempted solutions to some of these problems see Essays 6–10 of *Essays on Actions and Events*, and Essays 6–8 of this book. There is further discussion in Essays 3, 4, 9, and 10, and reference to some progress in section 1 of Essay 9.

3 *True to the Facts*

A true statement is a statement that is true to the facts. This remark seems to embody the same sort of obvious and essential wisdom about truth as the following about motherhood: a mother is a person who is the mother of someone. The *property* of being a mother is explained by the *relation* between a woman and her child; similarly, the suggestion runs, the property of being true is to be explained by a relation between a statement and something else. Without prejudice to the question what the something else might be, or what word or phrase best expresses the relation (of being true to, corresponding to, picturing), I shall take the licence of calling any view of this kind a *correspondence theory* of truth.

Correspondence theories rest on what appears to be an ineluctable if simple idea, but they have not done well under examination. The chief difficulty is in finding a notion of fact that explains anything, that does not lapse, when spelled out, into the trivial or the empty. Recent discussion is thus mainly concerned with deciding whether some form of correspondence is true and trivial ('... the theory of truth is a series of truisms'[1]) or, in so far as it is not confused, simply empty ('The correspondence theory requires, not purification, but elimination'[2]). Those who have discussed the semantic concept of truth in connection with correspondence theories have typically ruled the semantic concept either irrelevant or trivial.

In this paper I defend a version of the correspondence theory. I think truth can be explained by appeal to a relation between language and the world, and that analysis of that relation yields insight into how, by uttering sentences, we sometimes manage to say

[1, 2] J. L. Austin and P. F. Strawson, symposium on 'Truth'. The quoted remarks are from Austin and Strawson respectively, and appear on pages 115 and 129.

what is true. The semantic concept of truth, as first systematically expounded by Tarski,[3] will play a crucial role in the defence.

It might be possible to prove that any theory or definition of truth meeting plausible standards necessarily contains conceptual resources adequate to define a sense of correspondence. My project is less ambitious: I shall be satisfied if I can find a natural interpretation of the relation of correspondence that helps explain truth. Clearly it is consistent with the success of this attempt that there be a formula for eliminating phrases like 'it is true that' and 'is true' from many or all contexts: correspondence and redundancy theories do not necessarily conflict. Nevertheless, we may find instruction concerning the role of correspondence by asking how well we can do in systematically replacing sentences with truth-words or phrases by sentences without.

The sentence

(1) The statement that French is the official language of Mauritius is true.

is materially equivalent to 'French is the official language of Mauritius'; and the same might be said for any two sentences similarly related. This encourages the thought that the words that bed the embedded sentence in (1) represent an identity truth function, the same in power as double negation, but lacking significant articulation. On this suggestion, it would be no more than a freak of grammar that (1) consists of a complex singular term and a predicate.

The trouble with the double-negation theory of truth is that it applies only to sentences that have embedded sentences, like (1) and 'It is true that $2 + 2 = 5$'. The theory cannot cope directly with

(2) The Pythagorean theorem is true.
(3) Nothing Aristotle said is true.

We might retain the double-negation theory as applied to (1) while reserving (2) and (3) for separate treatment. But it is hardly plausible that the words 'is true' have different meanings in these different cases, especially since there seem to be simple inferences connecting sentences of the two sorts. Thus from (2) and 'The Pythagorean theorem is the statement that the square of the hypotenuse is equal

[3] A. Tarski, 'The Concept of Truth in Formalized Languages'.

to the sum of the squares on the other two sides' we can infer 'The statement that the square on the hypotenuse is equal to the sum of the squares on the other two sides is true'.

It is tempting to think that the double-negation theory can somehow be extended to cover sentences like (2) and (3). The reasoning might go this way: the double-negation theory tells us that for each statement there is a sentence that expresses it. But then (2) holds just in case there is a true sentence that expresses the Pythagorean theorem, and (3) holds in case no true sentence expresses something Aristotle said. The seeming need, in this explanation, to use the word 'true' will be shown to be harmless by rendering (2) and (3) thus:

(2′) (p) (the statement that $p =$ the Pythagorean theorem $\rightarrow p$)
(3′) $\sim(\exists p)$(Aristotle said that $p \cdot p$)

We are now pursuing a line that diverges from the simple double-negation theory by accepting an ontology of statements, and by introducing quantification into positions that can be occupied by sentences. Not that the variables in (2′) and (3′) range over statements; it is rather expressions of the sort flanking the identity sign in (2′) that refer to statements. In the double-negation theory, putative reference to statements and putative predication of truth were absorbed into a grammatically complex, but logically simple, expression, a truth-functional sentential connective. By contrast, the present theory allows us to view 'is true' as a genuine predicate. It provides a principle, namely

(4) (p)(the statement that p is true $\leftrightarrow p$)

that leads to sentences free of the predicate 'is true' and logically equivalent to sentences containing it. Here, truth is not explained away as something that can be predicated of statements, but explained.

Explained, that is, if we understand (2′), (3′), and (4). But do we? The trouble is in the variables. Since the variables replace sentences both as they feature after words like 'Aristotle said that' and in truth-functional contexts, the range of the variables must be entities that sentences may be construed as naming in both such uses. But there are very strong reasons, as Frege pointed out, for supposing that if sentences, when standing alone or in truth-functional contexts, name anything, then all true sentences name the same thing.[4] This would force us to conclude that the statement that *p* is identical with the statement that *q* whenever *p* and *q* are both true; presumably an unacceptable result.

In a brief, and often mentioned, passage F. P. Ramsey puts forward a theory similar to, or identical with, the one just discussed. He observes that sentences like (2′), (3′), and (4) cannot be convincingly read in English without introducing the words 'is true' at the end, but seems to see this as a quirk, or even defect, of the language (we add 'is true' because we forget '*p*' already contains a 'variable' verb). Ramsey then says

> This may perhaps be made clearer by supposing for a moment that only one form of proposition is in question, say the relational form *aRb*; then 'He is always right' could be expressed by 'For all *a*, *R*, *b*, if he asserts *aRb*, then *aRb*', to which 'is true' would be an obviously superfluous addition. When all forms of propositions are included the analysis is more complicated but not essentially different.[5]

I think we must assume that Ramsey wants the variables '*a*' and '*b*' to range over individuals of some sort, and '*R*' over (two-place) relations. So his version of 'He is always right' would be more fully expressed by 'For all *a*, *R*, *b*, if he asserts that *a* has *R* to *b*, then *a* has *R* to *b*'. Clearly, if 'all forms of proposition' are included, the analysis must be recursive in character, for the forms of propositions follow the (logical) forms of sentences, and of these there are an infinite number. There is no reason to suppose, then, that Ramsey's analysis could be completed in a way that did not essentially parallel Tarski's method for defining truth. Tarski's method, however, introduces (as I shall argue) something like the notion of correspondence, and this is just what the theories we have been exploring were supposed to avoid. Paradox may also be a problem for Ramsey's recursive project. Where a theory based on the principle of (4) can always informally plead that a term of the form 'the statement that *p*' fails to name when a troublesome sentence replaces '*p*', a theory that runs systematically through the sentences of a language will need to appeal to a more mechanical device to avoid contradiction. One wonders what conviction Ramsey's claim that 'there really is no separate problem of truth' would carry after his analysis was carried to completion.

[4] G. Frege, 'On Sense and Reference'. For the argument, see Essay 2.

[5] F. P. Ramsey, 'Facts and Propositions', 143.

I have said nothing about the purposes served in (non-philosophical) conversation by uttering sentences containing 'true' and cognates. No doubt the idea that remarks containing the word 'true' typically are used to express agreement, to emphasize conviction or authority, to save repetition, or to shift responsibility, would gain support if it could be shown that truth-words can always be eliminated without cognitive loss by application of a simple formula. Nevertheless, I would hold that theories about the extra-linguistic aims with which sentences are issued are logically independent of the question what they mean; and it is the latter with which I am concerned.

We have failed to find a satisfactory theory to back the thesis that attributions of truth to statements are redundant; but even if it could be shown that no such theory is possible, this would not suffice to establish the correspondence theory. So let us consider more directly the prospects for an account of truth in terms of correspondence.

It is facts correspondence to which is said to make statements true. It is natural, then, to turn to talk of facts for help. Not much can be learned from sentences like

(5) The statement that Thika is in Kenya corresponds to the facts.

or such variants as 'It is a fact that Thika is in Kenya', 'That Thika is in Kenya is a fact', and 'Thika is in Kenya, and that's a fact'. Whether or not we accept the view that correspondence to facts explains truth, (5) and its kin say no more than 'The statement that Thika is in Kenya is true' (or 'Is it true that . . .' or '. . . , and that's the truth', etc.). If (5) is to take on independent interest, it will be because we are able to give an account of facts and correspondence that does not circle back immediately to truth. Such an account would enable us to make sense of sentences with this form:

(6) The statement that *p* corresponds to the fact that *q*.

The step to truth would be simple: a statement is true if there is a fact to which it corresponds. [(5) could be rewritten 'The statement that Thika is in Kenya corresponds to a fact'.]

When does (6) hold? Certainly when '*p*' and '*q*' are replaced by the same sentence; after that the difficulties set in. The statement that Naples is farther north than Red Bluff corresponds to the fact that Naples is farther north than Red Bluff, but also, it would seem, to the fact that Red Bluff is farther south than Naples (perhaps these are the same fact). Also to the fact that Red Bluff is farther south than the largest Italian city within thirty miles of Ischia. When we reflect that Naples is the city that satisfies the following description: it is the largest city within thirty miles of Ischia, and such that London is in England, then we begin to suspect that if a statement corresponds to one fact, it corresponds to all. ('Corresponds to the *facts*' may be right in the end). Indeed, employing principles implicit in our examples, it is easy to confirm the suspicion. The principles are these: if a statement corresponds to the fact described by an expression of the form 'the fact that *p*', then it corresponds to the fact described by 'the fact that *q*' provided either (1) the sentences that replace '*p*' and '*q*' are logically equivalent, or (2) '*p*' differs from '*q*' only in that a singular term has been replaced by a coextensive singular term. The confirming argument is this. Let '*s*' abbreviate some true sentence. Then surely the statement that *s* corresponds to the fact that *s*. But we may substitute for the second '*s*' the logically equivalent '(the *x* such that *x* is identical with Diogenes and *s*) is identical with (the *x* such that *x* is identical with Diogenes)'. Applying the principle that we may substitute coextensive singular terms, we can substitute '*t*' for '*s*' in the last quoted sentence, provided '*t*' is true. Finally, reversing the first step we conclude that the statement that *s* corresponds to the fact that *t*, where '*s*' and '*t*' are any true sentences.[6]

Since aside from matters of correspondence no way of distinguishing facts has been proposed, and this test fails to uncover a single difference, we may read the result of our argument as showing that there is exactly one fact. Descriptions like 'the fact that there are stupas in Nepal', if they describe at all, describe the same thing: The Great Fact. No point remains in distinguishing among various names of The Great Fact when written after 'corresponds to'; we may as well settle for the single phrase 'corresponds to The Great Fact'. This unalterable predicate carries with it a redundant whiff of ontology, but beyond this there is apparently no telling it apart from 'is true'.

The argument that led to this conclusion could be thwarted by refusing to accept the principles on which it was based. And one can

[6] See footnote 4 above, and footnote 3 of Essay 2. For further discussion of the argument, and some surprising applications, see J. Wallace, 'Propositional Attitudes and Identity'.

certainly imagine constructing facts in ways that might reflect some of our feeling for the problem without leading to ontological collapse. From the point of view of the theory of truth, however, all such constructions seem doomed by the following difficulty. Suppose, to leave the frying-pan of extensionality for the fires of intension, we distinguish facts as finely as statements. Of course, not every statement has its fact; only the true ones do. But then, unless we find another way to pick out facts, we cannot hope to explain truth by appeal to them.[7]

Talk about facts reduces to predication of truth in the contexts we have considered; this might be called the *redundancy theory* of facts. Predications of truth, on the other hand, have not proved so easy to eliminate. If there is no comfort for redundancy theories of truth in this, neither is there encouragement for correspondence theories.

I think there is a fairly simple explanation for our frustration: we have so far left language out of account. Statements are true or false because of the words used in making them, and it is words that have interesting, detailed, conventional connections with the world. Any serious theory of truth must therefore deal with these connections, and it is here if anywhere that the notion of correspondence can find some purchase. We have been restricting ourselves to ways of specifying statements that make no apparent mention of words. Thus 'Jones's statement that the cat is on the mat' irretrievably washes out reference to the particulars of Jones's language that might support a non-trivial account of truth, and the same may be thought to hold for the 'the statement that *p*' idiom generally.

Discussions of truth may have avoided the linguistic turn because it is obvious that truth cannot be pinned on sentences; but if this has been a motive, it is a confused one. Sentences cannot be true or false because if they were we should have to say that 'Je suis Titania' was true (spoken or sung by Titania), false (spoken by anyone else), and neither (uttered by someone with no French). What this familiar argument shows is not that we must stop talking of sentences and truth in the same breath, but that we must breathe a little deeper and talk also of the time the sentence is uttered, and its utterer. Truth (in a given natural language) is not a property of sentences; it is a relation between sentences, speakers, and dates. To view it thus is not to turn away from language to speechless eternal entities like

[7] A similar point is made by P. F. Strawson, 'Truth: A Reconsideration of Austin's Views'.

propositions, statements, and assertions, but to relate language with the occasions of truth in a way that invites the construction of a theory.

The last two paragraphs may suggest that if we are to have a competent theory about truth we must abandon the view that statements are the proper vehicles of truth. But this is not so. If I am right, theories of truth must characterize or define a three-place predicate 'T *s*,*u*,*t*'. It will not matter to the theory whether we read this predicate 'sentence *s* is true (as English) for speaker *u* at time *t*' or 'the statement expressed by sentence *s* (as English) by speaker *u* at *t* is true'. Those who believe we must, for further reasons, retain statements as truth vehicles will find the second formulation, with its complex singular term ('the statement . . .') and one-place predicate ('is true') more perspicuous, while those who (with me) think we can get along without statements may prefer the more austere first formulation. But either party may talk either way; the difference comes out only when the talk is seen in the light of a comprehensive theory. Whether that theory requires an ontology of statements is not settled, I think, by the matters under discussion.

There are excellent reasons for not predicating truth of sentences, but these reasons do not apply to speech acts, utterances, or tokens. It has been argued, and convincingly, that we do not generally, or perhaps ever, say of a speech act, utterance, or token, that it is true.[8] This hardly shows why we *ought* not to call these entities (if they exist) true. No confusion would result if we said that the particular speaking of a sentence was true just in case it was used on that occasion to make a true statement; and similarly for tokens and utterances. According to Strawson,

> 'My statement' may be either what I say or my saying it. My saying something is certainly an episode. What I say is not. It is the latter, not the former, we declare to be true.[9]

I'm not sure a statement is ever a speech act, but in any case we may accept the conclusion that speech acts are not said to be true. But what follows? Certainly not that we cannot explain what it is to make a true statement in terms of the conventional relations between words and things that hold when the words are used by particular agents on particular occasions. For although 'my state-

[8] See R. Cartwright, 'Propositions'.
[9] P. F. Strawson, 'Truth', 129–30.

ment' may not refer, at least when truth is in question, to a speech act, still it may succeed in identifying its statement only by relating it to a speech act. (What makes it 'my' statement?)

If someone speaking English utters the sentence 'The sun is over the yard-arm', under what conditions has he made the statement that the sun is over the yard-arm? One range of answers might include such provisions as that he intended to convey to his hearers the impression that he believed the sun was over the yard-arm, that he was authorized by his status to issue information about the location of the sun, etc. Thinking along these lines, one might maintain that, if the speaker had no thought of the location of the sun, and wanted to announce that it was time for a drink, then he *didn't* make the statement that the sun is over the yard-arm. But there is also a sense of making a statement in which we would say, even under conditions of the sort just mentioned, that the man had ('literally') made the statement that the sun was over the yard-arm, and that what he said was ('literally') true provided the sun was over the yard-arm at the time he spoke, even though he had no reason to believe it, and didn't care if it were true. In such cases, we are interested not in what the person meant by uttering the sentence, but what the sentence, as uttered, meant. Both of these notions of meaning are relative to the circumstances of performance, but in the second case we abstract away from the extra-linguistic intentions of the speaker. Communication by language is communication by way of literal meaning; so there must be the literal sense of making a statement if there are others. The theory of truth deals with the literal sense. (Of course this point deserves to be discussed at much greater length.)

Cleaving to the literal, then, someone speaking English will make a true statement by uttering the sentence 'It's Tuesday' if and only if it is Tuesday in his vicinity at the time he speaks. The example invites generalization: every instance of the following schema will be a truth about truth when 's' is replaced by a description of a sentence of English and 'p' is replaced by a sentence that gives the conditions under which the described sentence is true:

(7) Sentence s is true (as English) for speaker u at time t if and only if p.[10]

[10] The expression that replaces 'p' must contain 'u' and 't' as free variables unless there are no indexical elements in s.

(An alternative schema apparently attributing truth to statements could be substituted.) Even if we restrict the descriptions we substitute for '*s*' to some stylized vocabulary of syntax, we may assume that there is a true sentence of the form of (7) for each English sentence. The totality of such sentences uniquely determines the extension of the three-place predicate of (7) (the relativized truth predicate). We seem here on the verge of a theory of truth; yet nothing like correspondence is in sight. The reason may be, however, that we are *only* on the verge of a theory. Schema (7) tells us what a theory of truth should encompass, but it is not such a theory itself, and does not suggest how such a theory can be contrived. Schema (7) is meant to play for English a role analogous to that played for an artificial language by a similar schema in Tarski's convention T.[11] The role is that of providing a test of the adequacy of a theory of truth: an acceptable theory must entail a true sentence with the form of (7) no matter what sentence of English is described by the canonical expression that replaces '*s*'.

Schema (7) lacks an elegant feature of its analogue in the work of Tarski. Tarski, not concerned with languages with indexical elements, can use this simple formula: '*s* is true (in L) if and only if *p*' where the sentence substituted for '*p*' is the sentence described by the expression that replaces '*s*' if the metalanguage contains the object language; otherwise it translates that sentence in some straightforward sense. This uncomplicated formula cannot be ours; for when there are indexical terms (demonstratives, tenses), what goes for '*p*' cannot in general be what '*s*' names or a translation of it, as witness the example in the first sentence of the preceding paragraph. The elaboration called for to state (7) in explicit syntactic terms would be considerable, but there is no reason to think it impossible since what replaces '*p*' must be systematically related to the sentence described by the replacement of '*s*' by the rules that govern the use of indexical terms in English.

If the indicative sentences of English comprised just a finite number of elementary sentences and truth-functional compounds of them, it would be easy to give a recursive characterization of truth by providing a sentence with the form of (7) for each elementary sentence, and giving a rule corresponding to each sentential connective. This strategy breaks down, however, as soon as we allow

[11] A. Tarski, 'The Concept of Truth in Formalized Languages', 187, 188.

predicates of arbitrary complexity to be built up using variables and connectives, as in quantification or complex singular terms; and it is just here that the theory of truth becomes interesting. Let us concentrate on quantificational structure at the expense of singular terms, not only because the latter are arguably dispensable while the former is not, but also because the point to be made will come through more simply. The problem presented by quantificational structure for a recursive theory of truth is, of course, that although sentences of any finite length can be constructed from a small supply of variables, connectives, predicates, and quantifiers, none of the parts of a sentence needs to be a sentence in turn; therefore the truth of a complex sentence cannot in general be accounted for in terms of the truth of its parts.

Tarski taught us to appreciate the problem and he gave an ingenious solution. The solution depends on first characterizing a relation called *satisfaction* and then defining truth by means of it. The entities that are satisfied are sentences both open and closed; the satisfiers are functions that map the variables of the object language on to the entities over which they range—almost everything, if the language is English.[12] A function satisfies an unstructured *n*-place predicate with variables in its *n* places if the predicate is *true of* the entities (in order) that the function assigns to those variables. So if '*x* loves *y*' is an open sentence of the simplest kind, a function *f* satisfies it just in case the entity that *f* assigns to '*x*' loves the entity that *f* assigns to '*y*'. The recursive characterization of satisfaction must run through every primitive predicate in turn. It copes with connectives in the obvious way: thus a conjunction of two sentences *s* and *t* (open or closed) is satisfied by *f* provided *f* satisfies *s* and *f* satisfies *t*. The universal quantification of an open sentence *s* with respect to a variable *v* is satisfied by *f* in case *f*, and every other function like *f* except in what it assigns to *v*, satisfies *s*. (The previous sentence works with 'existential' replacing 'universal' and 'or some' replacing 'and every'.) Whether or not a particular function satisfies a sentence depends entirely on what entities it assigns to the free variables of the sentence. So if the sentence has no free variables—if it is a closed, or genuine, sentence—then it must be satisfied by every function or by none. And, as is clear from the details of the recursion, those closed

[12] Tarski's satisfiers are infinite sequences, not functions. The reader in search of precision and deeper understanding cannot be too strongly urged to study 'The Concept of Truth in Formalized Languages'.

sentences which are satisfied by all functions are true; those which are satisfied by none are false. (I assume throughout that satisfaction, like truth, is relativized in the style of (7).)

The semantic concept of truth as developed by Tarski deserves to be called a correspondence theory because of the part played by the concept of satisfaction; for clearly what has been done is that the property of being true has been explained, and non-trivially, in terms of a relation between language and something else. The relation, satisfaction, is not, it must be allowed, exactly what intuition expected of correspondence; and the functions or sequences that satisfy may not seem much like facts. In part the contrast is due to a special feature of variables: just because they refer to no particular individual, satisfaction must consider arbitrary assignments of entities to variables (our functions). If we thought of proper names instead, satisfiers could be more nearly the ordinary objects of our talk—namely, ordered *n*-tuples of such. Thus 'Dolores loves Dagmar' would be satisfied by Dolores and Dagmar (in that order), provided Dolores loved Dagmar. I suppose Dolores and Dagmar (in that order) is not a fact either—the fact that verifies 'Dolores loves Dagmar' should somehow include the loving. This 'somehow' has always been the nemesis of theories of truth based on facts. So the present point isn't that '*s* is satisfied by all functions' means exactly what we thought '*s* corresponds to the facts' meant, only that the two phrases have this in common: both intend to express a relation between language and the world, and both are equivalent to '*s* is true' when *s* is a (closed) sentence.

The comparison between correspondence theories that exploit the concept of satisfaction and those which rest on correspondence to facts is at its best with sentences without free variables. The parallel even extends, if we accept Frege's argument about the extensions of sentences, to the conclusion that true sentences cannot be told apart in point of what they correspond to (the facts, The Great Fact) or are satisfied by (all functions, sequences). But Tarski's strategy can afford this sameness in the finished product where the strategy of facts cannot, because satisfaction of closed sentences is explained in terms of satisfaction of sentences both open and closed, whereas it is only closed sentences that traditionally have corresponding facts. Since different assignments of entities to variables satisfy different open sentences and since closed sentences are constructed from open, truth is reached, in the semantic approach, by different routes for different sentences. All true sentences end up in the same place, but there are different stories about how they got there; a semantic theory of truth tells the story for a particular sentence by running through the steps of the recursive account of satisfaction appropriate to the sentence. And the story constitutes a proof of a theorem in the form of an instance of schema (7).

The strategy of facts can provide no such instructive variety. Since all true sentences have the same relation to the facts, an explanation of the truth of a sentence on the basis of its relations to other (closed) sentences must, if it sticks to the facts, begin where it ends.

Seen in retrospect, the failure of correspondence theories of truth based on the notion of fact traces back to a common source: the desire to include in the entity to which a true sentence corresponds not only the objects the sentence is 'about' (another idea full of trouble) but also whatever it is the sentence says about them. One well-explored consequence is that it becomes difficult to describe the fact that verifies a sentence except by using that sentence itself. The other consequence is that the relation of correspondence (or 'picturing') seems to have direct application to only the simplest sentences ('Dolores loves Dagmar'). This prompts fact-theorists to try to explain the truth of all sentences in terms of the truth of the simplest and hence in particular to interpret quantification as mere shorthand for conjunctions or alternations (perhaps infinite in length) of the simplest sentences. The irony is that, in so far as we can see quantification in this light, there is no real need for anything like correspondence. It is only when we are forced to take generality as an essential addition to the conceptual resources of predication and the compounding of sentences, and not reducible to them, that we appreciate the uses of a sophisticated correspondence theory. Theory of truth based on satisfaction is instructive partly because it is less ambitious about what it packs into the entities to which sentences correspond: in such a theory, these entities are no more than arbitrary pairings of the objects over which the variables of the language range with those variables. Relative simplicity in the objects is offset by the trouble it takes to explain the relation between them and sentences, for every truth-relevant feature of every sentence must be taken into account in describing satisfaction. The pay-off is clear: in explaining truth in terms of satisfaction, all the conceptual resources of the language in relation to its ontology are brought to bear.

Talk of sentences', or better, statements', being true to, or corresponding to, the facts is of course as harmless as talk of truth. Even the suggestion in these phrases that truth is owed to a relation between language and the world can, I have argued, be justified. The strategy of facts, against which I have just been inveighing, is something else: a philosophical theory, and a bad one. It would be a shame to discredit all correspondence theories, and in particular Tarski's semantical approach, through thinking they must share the inadequacies of the usual attempts to explain truth on the basis of facts.

The assumption that all correspondence theories must use the strategy of facts is at least understandable and, given the vagaries of philosophical usage, could be considered true by fiat. There is less excuse for the widespread misunderstanding of the role of formulas like (7) in the semantical approach. The following example is no worse than many that could be quoted:

> . . . unless there is more to the 'correspondence' insisted on by classical correspondence theories of truth than is captured by the formulations of current semantic theory and unless this more can be shown to be an essential property of truth (or, at least, of a significant variety of truths), then the battle over correspondence, instead of being *won* by correspondence theorists, has shown itself to be a *Scheinstreit*. For, as has often been noted, the formula
>
> 'Snow is white' (in our language) is true = Snow is white
>
> is viewed with the greatest equanimity by pragmatist and coherentist alike. If the 'correspondence' of the correspondence theorist amounts to nothing more than is illustrated by such equivalences, then, while pragmatist and coherentist may hope to make important points, . . . nothing further would remain to be said about 'truth and correspondence'.[13]

Whether or not there is more to the semantic approach to truth than Sellars is ready to allow, it may be the case that no battle is won, or even joined, between correspondence theories and others. My trouble with this passage hinges on its assumption that a sentence like '"Snow is white" is true if and only if snow is white' (even when properly relativized and with a structural description in place of the quotation) in itself provides a clue to what is unique to the semantical approach. Of course, as Sellars says, such sentences are neutral ground; it is just for this reason that Tarski hopes everyone can agree that an adequate theory or definition of truth

[13] W. Sellars, 'Truth and "Correspondence"', 29.

must entail all sentences of this form. There is no trace of the notion of correspondence in these sentences, no relational predicate that expresses a relation between sentences and what they are about. Where such a relation, satisfaction, *does* come into play is in the elaboration of a non-trivial theory capable of meeting the test of entailing all those neutral snow-bound trivialities.

I would like now and by way of conclusion to mention briefly two of the many kinds of obstacle that must be overcome if we are to have a comprehensive theory of truth for a natural language. First, it is certainly reasonable to wonder to what extent it will ever be possible to treat a natural language as a formal system, and even more to question whether the resources of the semantical method can begin to encompass such common phenomena as adverbial modification, attributive adjectives, talk of propositional attitudes, of causality, of obligation, and all the rest. At present we do not even have a satisfactory semantics for singular terms, and on this matter many others hang. Still, a degree of optimism is justified. Until Frege, serious semantics was largely limited to predication and the truth-functional compounding of sentences. By abstracting quantificational structure from what had seemed a jungle of pronouns, quantifiers, connectives, and articles, Frege showed how an astonishingly powerful fragment of natural language could be semantically tamed. Indeed, it may still turn out that this fragment will prove, with ingenuity, to be the whole. Meanwhile, promising work goes on in many directions, enlarging the resources of formal semantics, extending the application of known resources, and providing the complex and detailed rules necessary to give a revealing description of the structure of natural language. Whatever range the semantic theory of truth ultimately turns out to have, we may welcome the insight that comes where we understand language well enough to apply it.

The second difficulty is on another level: we have suggested how it might be possible to interpret attributions of truth to statements or to sentences relativized to occasions of use, but only in contexts of the sort provided by the left branch of (7). We have given no indication of how the analysis could be extended to apply to sentences like

(8) It is true that it is raining.
(9) The statement that it is raining is true.

Here is how we might try to meet the case of (8). We have, we are supposing, a theory of truth-in-English with truth treated as a relation between a sentence, a speaker, and a time. (The alternative version in terms of statements would apply to (9).) The problem is to find natural counterparts of these elements in (8). A speaker of (8) speaks the words 'it is raining', thus performing an act that embodies a particular sentence, has its speaker, and its time. A reference to this act can therefore serve as a reference to the three items needed to apply the theory of truth. The reference we can think of as having been boiled down into the demonstrative 'that' of (8) and (9). A long-winded version of (8) might, then, go like this. First (reversing the order for clarity) I say 'It's raining'. Then I say '*That* speech act embodied a sentence which, spoken by me now, is true'. On this analysis, an utterance of (8) or (9) consists of two logically (semantically) independent speech acts, one of which contains a demonstrative reference to the other. An interesting feature of these utterances is that one is true if the other is; perhaps this confirms an insight of the redundancy theory.

A further problem is raised by

(10) Peter's statement that Paul is hirsute is true.

Following the suggestion made for (8) and (9), the analysis of (10) should be 'Paul is hirsute. That is true, and Peter said (stated) it'. The 'that', as before, refers to an act of speaking, and now the 'it' picks up the same reference. What is needed to complete the account is a paratactic analysis of indirect discourse that interprets an utterance by a speaker *u* of 'Peter said that Paul is hirsute' as composed of an utterance of 'Paul is hirsute' and another utterance ('Peter said that') that relates Peter in a certain way to *u*'s utterance of 'Paul is hirsute'. The relation in question can, perhaps, be made intelligible by appeal to the notion of *samesaying*: if *u* says what is true when he says 'Peter said that', it is because, by saying 'Paul is hirsute', he has made Peter and himself samesayers.[14]

One may, of course, insist that the relation of samesaying (which holds between speech acts) can be understood only by reference to a third entity: a statement, meaning, or proposition. Nothing I have written here bears on this question, except indirectly, by showing that, with respect to the problems at hand, no need arises for such

[14] For more on this approach to indirect discourse, see Essay 7.

entities. Is this simply the result of neglecting troublesome cases? Consider, as a final example,

(11) Peter said something true.

This cannot be rendered, 'Some (past) utterance of Peter's makes us samesayers', for I may not have said, or know how to say, the appropriate thing. Nor will it help to try 'Some utterance of Peter's embodied a sentence true under the circumstances'. This fails because (11) does not tell what language Peter spoke, and the concept of truth with which we are dealing is necessarily limited to a specific language. Not knowing what his language is, we cannot make sense of 'true-in-his-language'.

What we can hope to make sense of, I think, is the idea of a sentence in another tongue being the *translation* of a sentence of English. Given this idea, it becomes natural to see (11) as meaning something like 'Peter uttered a sentence that translates a sentence of English true under the circumstances'. The exact nature of the counterfactual assumption barely concealed in this analysis depends on the details of the theory of truth (for English) as relativized to occasions of utterance. In any case, we seem required to understand what someone else would mean by a sentence of our language if he spoke our language. But difficult as this concept is, it is hard to see how communication can exist without it.

The conclusion I would tentatively draw is this. We can get away from what seems to be talk of the (absolute) truth of timeless statements if we accept truth as relativized to occasions of speech, and a strong notion of translation. The switch may create more problems than it solves. But they are, I think, the right problems: providing a detailed account of the semantics of natural language, and devising a theory of translation that does not depend upon, but rather founds, whatever there is to the concept of meaning.

Strawson describes Austin's 'purified version of the correspondence theory of truth' in this way:

> His . . . theory is, roughly, that to say that a statement is true is to say that a certain speech-episode is related in a certain conventional way to something in the world exclusive of itself.[15]

It is this theory Strawson has in mind when he says, 'The correspondence theory requires, not purification, but elimination'. I

[15] P. F. Strawson, 'Truth', 129.

would not want to defend the details of Austin's conception of correspondence, and many of the points I have made against the strategy of facts echo Strawson's criticisms. But the debilities of particular formulations of the correspondence theory ought not be held against the theory. If I am right, by appealing to Tarski's semantical conception of truth we can defend a theory that almost exactly fits Strawson's description of Austin's 'purified version of the correspondence theory of truth'. And this theory deserves, not elimination, but elaboration.

4 *Semantics for Natural Languages*

A theory of the semantics of a natural language aims to give the meaning of every meaningful expression, but it is a question what form a theory should take if it is to accomplish this. Since there seems to be no clear limit to the number of meaningful expressions, a workable theory must account for the meaning of each expression on the basis of the patterned exhibition of a finite number of features. Even if there is a practical constraint on the length of the sentences a person can send and receive with understanding, a satisfactory semantics needs to explain the contribution of repeatable features to the meaning of sentences in which they occur.

I suggest that a theory of truth for a language does, in a minimal but important respect, do what we want, that is, give the meanings of all independently meaningful expressions on the basis of an analysis of their structure. And on the other hand, a semantic theory of a natural language cannot be considered adequate unless it provides an account of the concept of truth for that language along the general lines proposed by Tarski for formalized languages. I think both linguists and philosophers interested in natural languages have missed the key importance of the theory of truth partly because they have not realized that a theory of truth gives a precise, profound, and testable answer to the question how finite resources suffice to explain the infinite semantic capacities of language, and partly because they have exaggerated the difficulties in the way of giving a formal theory of truth for a natural language. In any event the attempt is instructive, for in so far as we succeed in giving such a theory for a natural language, we see the natural language as a formal system; and in so far as we make the construction of such a theory our aim, we can think of linguists and analytic philosophers as co-workers'.[1]

By a theory of truth I mean a set of axioms that entail, for every sentence in the language, a statement of the conditions under which it is true. Obviously if we have a definition of a truth predicate satisfying Tarski's Convention T we have a theory of truth,[2] but in general the characterization of a theory of truth demands less. If no further restrictions are imposed, some theories of truth will be of little intrinsic interest. For example, we could simply take as axioms all sentences of the form '*s* is true if and only if *p*' where '*s*' is to be replaced by a standardized description of a sentence and '*p*' by that sentence (assuming that the metalanguage contains the object language). Such a theory would yield no insight into the structure of the language and would thus provide no hint of an answer to the question how the meaning of a sentence depends on its composition. We could block this particular aberration by stipulating that the non-logical axioms be finite in number; in what follows I shall assume that this restriction is in force, though it may be that other ways exist of ensuring that a theory of truth has the properties we want.

What properties do we want? An acceptable theory should, as we have said, account for the meaning (or conditions of truth) of every sentence by analysing it as composed, in truth-relevant ways, of elements drawn from a finite stock.[3] A second natural demand is that the theory provide a method for deciding, given an arbitrary sentence, what its meaning is. (By satisfying these two conditions a theory may be said to show that the language it describes is *learnable* and *scrutable*.) A third condition is that the statements of truth conditions for individual sentences entailed by the theory should, in some way yet to be made precise, draw upon the same concepts as the sentences whose truth conditions they state.[4]

Theories of the sort Tarski showed how to devise clearly enough

[1] See footnote 14 to Essay 2.

[2] A. Tarski, 'The Concept of Truth in Formalized Languages'.

[3] Truth conditions are not to be equated with meanings; at best we can say that by giving the truth conditions of a sentence we give its meaning. But this claim too needs clarification. For some needed qualifications, see Essays 9 and 12. [Note added in 1982.]

[4] For important reservations on the possibility of making this condition precise, see S. Kripke, 'Is There a Problem about Substitutional Quantification?'. [Note added in 1982.]

have these desirable characteristics. The last condition, for example, is satisfied in an elementary way by a theory couched in a metalanguage that contains the object language, for in the required statements of the form '*s* is true if and only if *p*' the truth conditions of *s* are given by the sentence that replaces '*p*', namely *s* itself, and so make no use of any concepts not directly called upon in understanding *s*. If the metalanguage does not contain the object language, it is less obvious when this criterion is satisfied: and natural languages raise further problems that will be touched upon presently.

It seems natural to interpret the third condition as prohibiting the appearance of a semantic term in the statement of the truth conditions of a sentence unless that sentence already contains the semantic term (or a translation of it). It is not clear whether or not this would rule out explicit resort to semantic concepts in the statement of truth conditions for modal sentences (since it is uncertain whether or not these could be construed from the start as semantic in nature). But this constraint does appear to threaten theories that appeal to an unanalysed concept of denoting or naming, as well as those that make truth in a model the fundamental semantical notion.[5]

To put this interpretation on the third condition is, it seems, to judge much recent work in semantics irrelevant to present purposes; so I intend here to leave the question open, along with many further questions concerning the detailed formulation of the standards we should require of a theory of truth. My present interest is not in arguing disputed points but to urge the general relevance and productiveness of requiring of any theory of meaning (semantics) for a natural language that it give a recursive account of truth. It seems to me no inconsiderable merit of this suggestion that it provides a framework within which a multitude of issues and problems can be sharply stated.

To give a recursive theory of truth for a language is to show that the syntax of the language is formalizable in at least the sense that every true expression may be analysed as formed from elements (the 'vocabulary'), a finite supply of which suffice for the language by the application of rules, a finite number of which suffice for the language. But even if we go on to assume that falsehood may be defined in terms of truth, or independently and similarly characterized, it does not follow that sentencehood or grammaticalness can

[5] See J. Wallace, 'Nonstandard Theories of Truth'.

be recursively defined. So arguments designed to establish that a formal recursive account of syntax (sentencehood or grammaticalness) cannot be given for a natural language do not necessarily discredit the attempt to give a theory of truth. It should also be mentioned that the suggested conditions of adequacy for a theory of truth do not (obviously, anyway) entail that even the true sentences of the object language have the form of some standard logical system. Thus supposing it were clear (which it is not) that the deep structure of English (or another natural language) cannot be represented by a formal language with the usual quantificational structure, it still would not follow that there was no way of giving a theory of truth.

A theory of truth for a natural language must take account of the fact that many sentences vary in truth value depending on the time they are spoken, the speaker, and even, perhaps, the audience. We can accommodate this phenomenon either by declaring that it is particular utterances or speech acts, and not sentences, that have truth values, or by making truth a relation that holds between a sentence, a speaker, and a time.

To thus accommodate the indexical, or demonstrative, elements in a natural language is to accept a radical conceptual change in the way truth can be defined, as will be appreciated by reflection on how Convention T must be revised to make truth sensitive to context. But the change need not mean a departure from formality.

The fear is often expressed that a formal theory of truth cannot be made to cope with the problems of ambiguity in natural language that absorb so much of the energy of linguists. In thinking about this question it may help to distinguish two claims. One is that formal theories of truth have not traditionally been designed to deal with ambiguity, and it would change their character to equip them to do so. This claim is justified, and harmless. Theories of truth in Tarski's style do not in general treat questions of definition for the primitive vocabulary (as opposed to questions of translation and logical form); on the other hand there is nothing in a theory of truth inimical to the satisfactory treatment of the problems a lexicon is designed to solve. The second claim is that some sorts of ambiguity necessarily prevent our giving a theory of truth. Before this thesis can be evaluated it will be necessary to be far clearer than we are now about the criteria of success in giving a theory of truth for a natural language. Without attempting a deep discussion here, let me indicate why I think the issue cannot be settled by quoting a few puzzling cases.

Bar-Hillel gives an example like this: "They came by slow trains and plane.'[6] We can take 'slow' as modifying the conjunction that follows, or only 'train'. Of course an adequate theory would uncover the ambiguity; a theory of truth would in particular need to show how an utterance of the sentence could be true under one interpretation and false under another. So far there is no difficulty for a theory of truth. But Bar-Hillel makes the further observation that the context of utterance might easily resolve the ambiguity for any normal speaker of English, and yet the resolution could depend on general knowledge in a way that could not (practically, at least) be captured by a formal theory. By granting this, as I think we must, we accept a limitation on what a theory of truth can be expected to do. Within the limitation it may still be possible to give a theory that captures an important concept of meaning.

We have touched lightly on a few of the considerations that have led linguists and philosophers to doubt whether it is possible to give a formal theory of truth for a natural language. I have suggested that the pessimism is premature, particularly in the absence of a discussion of criteria of adequacy. On the other hand, it would be foolish not to recognize a difference in the interests and methods of those who study contrived languages and those who study natural languages.

When logicians and philosophers of language express reservations concerning the treatment of natural languages as formal systems, it may be because they are interested mainly in metatheoretical matters like consistency, completeness, and decidability. Such studies assume exact knowledge of the language being studied, a kind of precision that can be justified only by viewing the relevant features of the object language as fixed by legislation. This attitude is clearly not appropriate to the empirical study of language.

It would be misleading, however, to conclude that there are two kinds of language, natural and artificial. The contrast is better drawn in terms of guiding interests. We can ask for a description of the structure of a natural language: the answer must be an empirical theory, open to test and subject to error, and doomed to be to some extent incomplete and schematic. Or we can ask about the formal

[6] Y. Bar-Hillel, *Language and Information*, 182.

properties of the structures we thus abstract. The difference is like that between applied and pure geometry.

I have been urging that no definite obstacle stands in the way of giving a formal theory of truth for a natural language; it remains to say why it would be desirable. The reasons are of necessity general and programmatic, for what is being recommended is not a particular theory, but a criterion of theories. The claim is that if this criterion is accepted, the empirical study of language will gain in clarity and significance. The question whether a theory is correct can be made reasonably sharp and testable; the theories that are called for are powerful in explanatory and predictive power, and make use of sophisticated conceptual resources that are already well understood. Among the problems which a satisfactory theory of truth would solve, or help solve, are many which interest both linguists and philosophers; so as a fringe benefit we may anticipate a degree of convergence in the methods and interests of philosophy and linguistics. Let me try, briefly, to give colour to these remarks.

One relatively sharp demand on a theory for a language is that it give a recursive characterization of sentencehood. This part of theory is testable to the extent that we have, or can contrive, reliable ways of telling whether an expression is a sentence. Let us imagine for now that we can do this well enough to get ahead. In defining sentencehood what we capture, roughly, is the idea of an independently meaningful expression. But meaningfulness is only the shadow of meaning; a full-fledged theory should not merely ticket the meaningful expressions, but give their meanings. The point is acknowledged by many linguists today, but for the most part they admit they do not know how to meet this additional demand on theory, nor even how to formulate the demand.[7] So now I should like to say a bit more in support of the claim that a theory of truth does 'give the meaning' of sentences.

A theory of truth entails, for each sentence *s*, a statement of the form '*s* is true if and only if *p*' where in the simplest case '*p*' is replaced by *s*. Since the words 'is true if and only if' are invariant, we may interpret them if we please as meaning 'means that'. So construed, a sample might then read '"Socrates is wise" means that Socrates is wise'.

This way of bringing out the relevance of a theory of truth to

[7] For example see Chomsky's remarks on semantics, 'Topics in the Theory of Generative Grammar'.

questions of meaning is illuminating, but we must beware lest it encourage certain errors. One such error is to think that all we can learn from a theory of truth about the meaning of a particular sentence is contained in the biconditional demanded by Convention T. What we can learn is brought out rather in the *proof* of such a biconditional, for the proof must demonstrate, step by step, how the truth value of the sentence depends upon a recursively given structure. Once we have a theory, producing the required proof is easy enough; the process could be mechanized.

To see the structure of a sentence through the eyes of a theory of truth is to see it as built by devices a finite number of which suffice for every sentence; the structure of the sentence thus determines its relations to other sentences. And indeed there is no giving the truth conditions of all sentences without showing that some sentences are logical consequences of others; if we regard the structure revealed by a theory of truth as deep grammar, then grammar and logic must go hand in hand.

There is a sense, then, in which a theory of truth accounts for the role each sentence plays in the language in so far as that role depends on the sentence's being a potential bearer of truth or falsity; and the account is given in terms of structure. This remark is doubtless far less clear than the facts that inspire it, but my purpose in putting the matter this way is to justify the claim that a theory of truth shows how 'the meaning of each sentence depends on the meaning of the words'. Or perhaps it is enough to say that we have given a sense to a suggestive but vague claim; there is no reason not to welcome alternative readings if they are equally clear. In any case, to accept my proposal is to give up the attempt to find entities to serve as meanings of sentences and words. A theory of truth does without; but this should be counted in its favour, at least until someone gives a coherent and satisfactory theory of meaning that employs meanings.

Convention T, suitably modified to apply to a natural language, provides a criterion of success in giving an account of meaning. But how can we test such an account empirically? Here is the second case in which we might be misled by the remark that the biconditionals required by Convention T could be read as giving meanings, for what this wrongly suggests is that testing a theory of truth calls for direct insight into what each sentence means. But in fact, all that is needed is the ability to recognize when the required biconditionals are true. This means that in principle it is no harder to test the empirical adequacy of a theory of truth than it is for a competent speaker of English to decide whether sentences like '"Snow is white" is true if and only if snow is white' are true. So semantics, or the theory of truth at least, seems on as firm a footing empirically as syntax. It may in fact be easier in many cases for a speaker to say what the truth conditions of a sentence are than to say whether the sentence is grammatical. It may not be clear whether 'The child seems sleeping' is grammatical; but surely 'The child seems sleeping' is true if and only if the child seems sleeping.

I have been imagining the situation where the metalanguage contains the object language, so that we may ask a native speaker to react to the familiar biconditionals that connect a sentence and its description. A more radical case arises if we want to test a theory stated in our own language about the language of a foreign speaker. Here again a theory of truth can be tested, though not as easily or directly as before. The process will have to be something like that described by Quine in Chapter 2 of *Word and Object*. We will notice conditions under which the alien speaker assents to or dissents from, a variety of his sentences. The relevant conditions will be what we take to be the truth conditions of his sentences. We will have to assume that in simple or obvious cases most of his assents are to true, and his dissents from false, sentences—an inevitable assumption since the alternative is unintelligible. Yet Quine is right, I think, in holding that an important degree of indeterminacy will remain after all the evidence is in; a number of significantly different theories of truth will fit the evidence equally well.[8]

Making a systematic account of truth central in empirical semantics is in a way merely a matter of stating old goals more sharply. Still, the line between clarification and innovation in science is blurred, and it seems likely that the change would shift priorities in linguistic research. Some problems that have dominated recent work on semantics would fade in importance: the attempt to give 'the meaning' of sentences, and to account for synonymy, analyticity, and ambiguity. For the first of these, the theory of truth provides a kind of substitute; the second and third become unnecessary appendages; the fourth reappears in a special form. What would

[8] For further discussion of the principles guiding interpretation see Essays 2, 9–11, 13, 14. There are some differences between Quine's method of radical translation and the method I am suggesting; see Essay 16.

emerge as the deep problems are the difficulties of reference, of giving a satisfactory semantics for modal sentences, sentences about propositional attitudes, mass terms, adverbial modification, attributive adjectives, imperatives, and interrogatives; and so on through a long list familiar, for the most part, to philosophers.[9]

It is a question how much of a realignment we are talking about for linguistics. This depends largely on the extent to which the structure revealed by a theory of truth can be identified with the deep structure transformational grammarians seek. In one respect, logical structure (as we may call the structure developed by a theory of truth) and deep structure could be the same, for both are intended to be the foundation of semantics. Deep structure must also serve, however, as the basis for the transformations that produce surface structures, and it is an open question whether logical structure can do this job, or do it well.[10]

Finally, deep structure is asked by some linguists to reflect the 'internalized grammar' of speakers of the language. Chomsky in particular has argued that the superiority of transformational grammars over others that might be equally good at accounting for the totality of grammatical sentences lies in the fact that transformational grammars can be made to 'correspond to the linguistic intuition of the native speaker'.[11] The problem is to find a relatively clear test for when a theory corresponds to a speaker's linguistic intuition. I would like to suggest that we can give one sort of empirical bite to this idea if we take deep structure to be logical form. Consider this passage in Chomsky.[12]

Chomsky says that the following two sentences, though they have the same surface structure, differ in deep structure:

(1) I persuaded John to leave.
(2) I expected John to leave

The demonstration rests chiefly on the observation that when an embedded sentence in a sentence somewhat like (2) is transformed to the passive, the result is 'cognitively synonymous' with the active form; but a similar transformation does not yield a synonymous

[9] Since this was written, much has changed: linguists have recognized these problems and many more, while philosophers have been prodded by linguists to appreciate problems they had never noticed. [Note added in 1982.]

[10] For development of this theme, see G. Harman, 'Logical Form'.

[11] N. Chomsky, *Aspects of the Theory of Syntax*, 24.

[12] Ibid., 22.

result for the analogue of (1). The observation is clearly correct, but how does it show that (1) and (2) have radically different deep structures? At most the evidence suggests that a theory that assigns different structures to (1) and (2) may be simpler than one that does not. But how our linguistic intuitions have been tapped to prove a difference here is obscure.

But of course Chomsky is right; there is a contrast between (1) and (2), and it emerges dramatically the moment we start thinking in terms of constructing a theory of truth. Indeed we need go no further than to ask about the semantic role of the word 'John' in both sentences. In (1), 'John' can be replaced by any coreferring term without altering the truth value of (1); this is not true of (2). The contribution of the word 'John' to the truth conditions of (1) must therefore be radically different from its contribution to the truth conditions of (2). This way of showing there is a difference in the semantic structure of (1) and (2) requires no appeal to 'the speaker's tacit knowledge' of the grammar or the 'intrinsic competence of the idealized native speaker'. It rests on the explicit knowledge any speaker of English has of the way in which (1) and (2) may vary in truth under substitutions for the word 'John'.

Yet these last remarks do not begin to do justice to the method of truth. They show that by bearing the requirements of a theory of truth in mind we can throw into relief our feeling for a difference in structure between (1) and (2). So far, though, the evidence to which we are appealing is of much the same sort as Chomsky uses: mainly questions of the loss or preservation of truth value under transformations. Such considerations will no doubt continue to guide the constructive and analytic labours of linguists as they long have those of philosophers. The beauty of a theory of the sort we have been discussing is that these intimations of structure, however useful or essential they may be to the discovery of a suitable theory, need play no direct role in testing the final product.

5 *In Defence of Convention T*

Let someone say, 'There are a million stars out tonight' and another reply, 'That's true', then nothing could be plainer than that what the first has said is true if and only if what the other has said is true. This familiar effect is due to the reciprocity of two devices, one a way of referring to expressions (work done here by the demonstrative 'that'), the other the concept of truth. The first device takes us from talk of the world to talk of language; the second brings us back again.

We have learned to represent these facts by sentences of the form 'The sentence "There are a million stars out tonight" is true if and only if there are a million stars out tonight'. Because T-sentences (as we may call them) are so obviously true, some philosophers have thought that the concept of truth, at least as applied to sentences, was trivial. But of course this is wrong, as Ramsey was perhaps the first to point out, since T-sentences provide an alternative to talk of truth for some contexts only. T-sentences are no help if we want equivalents of 'Every sentence Aristotle spoke was false' or 'What you said last Tuesday was true'.[1]

T-sentences don't, then, show how to live without a truth predicate; but taken together, they do tell what it would be like to have one. For since there is a T-sentence corresponding to each sentence of the language for which truth is in question, the totality of T-sentences exactly fixes the extension, among the sentences, of any predicate that plays the role of the words 'is true'. From this it is clear that although T-sentences do not define truth, they can be used to define truth predicatehood: any predicate is a truth predicate that makes all T-sentences true.

[1] F. P. Ramsey, 'Facts and Propositions', 143.

T-sentences shed what light they do on truth by exhausting the cases, but since for any interesting language the number of cases is infinite, it is instructive to ask what properties truth must have if it is to act as the T-sentences say it does. To spell out the properties in an instructive, i.e. finite way, is to provide a theory of truth. We may think of a theory of truth for a language L simply as a sentence T containing a predicate *t* such that T has as logical consequences all sentences of the form '*s* is true if and only if *p*' with '*s*' replaced by a canonical description of a sentence of L, '*p*' replaced by that sentence (or its translation), and 'is true' replaced, if necessary, by *t*.

In essence this is, of course, Tarski's Convention T.[2] It differs in one significant respect only: it does not require that the conditions on the truth predicate suffice for an explicit definition. (I shall return to this point.) Throughout Tarski's writings on truth, Convention T provides the transition from informal intuitions concerning the concept of truth to sharp statement of problem. We first come across Convention T in 'The Concept of Truth in Formalized Languages' (published in Polish in 1933, in German in 1936, and in English in 1956). In a subsequent article aimed mainly at philosophers, Tarski appeals to T-sentences to relate classical philosophical issues to his work in semantics.[3] In a later popular piece he takes the same line.[4]

For all its familiarity, I think a central point about Convention T has been largely lost on philosophers. It is this: Convention T and T-sentences provide the sole link between intuitively obvious truths about truth and formal semantics. Without Convention T, we should have no reason to believe that truth is what Tarski has shown us how to characterize.

The weight Tarski puts on Convention T is interesting in another way, for it sets up a mild terminological strain. According to Tarski,

> *Semantics* is a discipline which, speaking loosely, deals with certain relations between expressions of a language and the objects . . . 'referred to' by those expressions.[5]

He gives as examples of semantic concepts *designation* (denotation), which relates singular terms and what they denote; *satisfaction*, which holds between an open sentence and the entity or entities it is

[2] A. Tarski, 'The Concept of Truth in Formalized Languages', 187, 188.
[3] A. Tarski, 'The Semantic Conception of Truth'.
[4] A. Tarski, 'Truth and Proof'.
[5] A. Tarski, 'The Semantic Conception of Truth', 345.

true of; and *definition*, which relates an equation and a number it uniquely determines. He continues,

While the words '*designates*', '*satisfies*', and '*defines*' express relations . . . , the word '*true*' is of a different logical nature: it expresses a property (or denotes a class) of certain expressions, viz., of sentences.[6]

He goes on to remark that 'the simplest and most natural way' of defining truth goes by way of satisfaction, and this explains his calling truth a semantic concept. But we cannot help noticing that truth is not, by Tarski's own account, a semantic concept. A glance at the logical grammar of T-sentences shows that it is essential that the truth predicate not express a relation: if it did, there could not be the crucial disappearance, from the right branch of the T-sentence biconditional, of all semantic concepts, and indeed of everything except the very sentence whose truth conditions it states (or a translation of that sentence).[7]

It would do much to justify Tarski's terminology if it could be shown that any theory that meets the criterion of Convention T has resources adequate to characterizing the satisfaction relation. Clearly there are languages for which this does not hold, but it would be interesting if for sufficiently rich languages it did. Recent work suggests a positive answer.[8] The feeling that truth belongs with the other semantic concepts is also enhanced by the fact that in each case there is a paradigm that both uses and mentions the same expression: 'Plato' designates Plato; y satisfies 'x is white' if and only if y is white; 'snow is white' is true if and only if snow is white; the equation '$2x = 1$' defines the number y if and only if $2y = 1$, etc. The existence of such a paradigm cannot, however, be taken as a sign of the semantical, for there is no appropriate analogous paradigm for relativized semantic concepts like truth in a model.

My present purpose is to defend, on grounds of philosophical interest and importance, concepts that meet the criterion of Convention T (and for which, then, there is a familiar paradigm).

[6] Ibid., 345.

[7] This claim is modified below.

[8] See J. Wallace, 'On the Frame of Reference' and 'Convention T and Substitutional Quantification'; also L. Tharp, 'Truth, Quantification, and Abstract Objects'. Although Kripke criticizes some of Wallace's and Tharp's arguments, he concurs with the conclusion that in a language with normal expressive power, some of the quantifiers at least must be provided with the kind of semantics for which Tarski's satisfaction is designed. See S. Kripke, 'Is There a Problem about Substitutional Quantification?'. [Footnote added in 1982.]

One reason Tarski appealed to Convention T was that he hoped to persuade philosophers that formal work in semantics was relevant to their concerns. It is a scandal that this cause still needs support in some quarters; but this is not my worry in this paper. I now address those who thoroughly appreciate the relevance of formal semantics to philosophical issues, but see no significant difference between theories that are in accord with Convention T, and theories that are not. (There is even a danger that the know-nothings and the experts will join forces; the former, hearing mutterings of possible worlds, trans-world heir lines, counterparts, and the like, are apt to think, *now* semantics is getting somewhere—out of this world, anyway.)

Before giving my reasons for thinking Convention T captures a concept worth attending to, I want to register some disclaimers.

First, to seek a theory that accords with Convention T is not, in itself at least, to settle for Model T logic or semantics. Convention T, in the skeletal form I have given it, makes no mention of extensionality, truth functionality, or first-order logic. It invites us to use whatever devices we can contrive appropriately to bridge the gap between sentence mentioned and sentence used. Restrictions on ontology, ideology, or inferential power find favour, from the present point of view, only if they result from adopting Convention T as a touchstone. What I want to defend is the Convention as a criterion of theories, not any particular theories that have been shown to satisfy the Convention in particular cases, or the resources to which they may have been limited.

Convention T defines a goal irrelevant to much contemporary work in semantics. Theories that characterize or define a relativized concept of truth (truth in a model, truth in an interpretation, valuation, or possible world) set out from the start in a direction different from that proposed by Convention T. Because they substitute a relational concept for the single-place truth predicate of T-sentences, such theories cannot carry through the last step of the recursion on truth or satisfaction which is essential to the quotation-lifting feature of T-sentences. There is a tradition, initiated by Tarski himself, of calling a relativized theory of truth the general theory of which the absolute theory (which satisfies Convention T) is a special case. Tarski says that 'the concept of *correct or true sentence in an individual domain* [is] a concept of a relative character [that] plays a much greater part than the absolute concept of truth and includes it as a special case'.[9] The sense in which this is correct is, of course, perfectly clear. On the other hand, it is important to remember that T-sentences do not fall out as theorems of a relativized theory of truth, and therefore such a theory does not necessarily have the same philosophical interest as a theory that satisfies Convention T.

The point of these remarks is not, I hope it is plain, to suggest that there is something wrong, defective, or misleading about semantic theories that do not yield T-sentences as theorems. On the contrary, it is obvious that such theories illuminate important concepts (such as completeness and logical consequence); I hardly need to expand on this theme in the present context. My thesis is only that there are important differences between theories of relative, and of absolute, truth, and the differences make theories of the two sorts appropriate as answers to different questions. In particular, philosophers of language have reason to be concerned with theories that satisfy Convention T.

Theories of truth based on the substitutional interpretation of quantification do not in general yield the T-sentences demanded by Convention T. There are exceptions in the case of object languages whose true atomic sentences can be effectively given, but the exceptions cannot include languages interestingly similar to natural languages in expressive power. Unlike theories of relative truth, substitutional theories have no evident virtue to set against their failure to satisfy Convention T.[10]

Finally, it is not my thesis that *all* we should demand of a semantic theory is that it meet the standard of Convention T. I shall suggest that theories satisfying Convention T are capable of explaining more than is often thought. But there may be choosing among such theories on the basis of further criteria; and of course there is much we want to know that lies outside.

Now for a catalogue of some of the features of theories that accord with Convention T and which should recommend such theories to philosophers of language.

The central merit of Convention T is that it substitutes for an important but murky problem a task whose aim is clear. After the substitution one appreciates better what was wanted in the first

[9] A. Tarski, 'The Concept of Truth in Formalized Languages', 199.

[10] While the general conclusion of these remarks may be right, the remarks are careless. In particular, I failed to consider languages with both substitutional and ontic quantifiers. (See footnote 8 above.) [Footnote added 1982.]

place, and gains insight into the aetiology of confusion. The original question is not confused, only vague. It is: what is it for a sentence (or utterance or statement) to be true? Confusion threatens when this question is reformulated as, what makes a sentence true? The real trouble comes when this in turn is taken to suggest that truth must be explained in terms of a relation between a sentence as a whole and some entity, perhaps a fact, or state of affairs. Convention T shows how to ask the original question without inviting these subsequent formulations. The form of T-sentences already hints that a theory can characterize the property of truth without having to find entities to which sentences that have the property differentially correspond. (I do not want to say that an absolute theory of truth is not in some sense a 'correspondence theory' of truth. But the entities it invokes are sequences, which are nothing like facts or states of affairs.[11])

A theory that satisfies Convention T has, then, the virtue of being an answer to a good question. The statement of the question is as persuasive, in relation to the intuitive concept of truth, as we find the truth of T-sentences to be. Both the statement of the question, and the character of the answers it allows, are revealing in that no semantic notions are used that are not fully characterized in relevant respects, syntactically by Convention T, materially by the theories it accepts.

A recursive theory of absolute truth, of the kind required by Convention T, provides an answer, *per accidens* it may at first seem, to quite another problem. This problem may be expressed as that of showing or explaining how the meaning of a sentence depends on the meanings of its parts. A theory of absolute truth gives an answer in the following sense. Since there is an infinity of T-sentences to be accounted for, the theory must work by selecting a finite number of truth-relevant expressions and a finite number of truth-affecting constructions from which all sentences are composed. The theory then gives outright the semantic properties of certain of the basic expressions, and tells how the constructions affect the semantic properties of the expressions on which they operate.

In the previous paragraph, the notion of meaning to which appeal is made in the slogan 'The meaning of the sentence depends on the meanings of its parts' is not, of course, the notion that opposes

[11] See Essay 3.

meaning to reference, or a notion that assumes that meanings are entities. The slogan reflects an important truth, one on which, I suggest, a theory of truth *confers* a clear content. That it does so without introducing meanings as entities is one of its rewarding qualities.

Theories of absolute truth necessarily provide an analysis of structure relevant to truth and to inference. Such theories therefore yield a non-trivial answer to the question what is to count as the logical form of a sentence. A theory of truth does not yield a definition of logical consequence or logical truth, but it will be evident from a theory of truth that certain sentences are true solely on the basis of the properties assigned to the logical constants. The logical constants may be identified as those iterative features of the language that require a recursive clause in the characterization of truth or satisfaction. Logical form, in this account, will of course be relative to the choice of a metalanguage (with its logic) and a theory of truth.

All that I have said so far in defence of Convention T is that it may be taken as clearly formulating some troublesome issues in the philosophy of language, and that if it is taken in this way, it invites solutions that have conspicuous merits. But why should we take it this way? May not some rival convention formulate the issues better? What follows is designed to bear on these questions.

The main, if not the only, ultimate concern of philosophy of language is the understanding of natural languages. There is much to be said for restricting the word 'language' to systems of signs that are or have been in actual use: uninterpreted formal systems are not languages through lack of meaning, while interpreted formal systems are best seen as extensions or fragments of the natural languages from which they borrow life.

The inevitable goal of semantic theory is a theory of a natural language couched in a natural language (the same or another). But as Tarski pointed out, on some obvious assumptions pursuit of this goal leads to paradox. He wrote:

> A characteristic feature of colloquial language . . . is its universality. It would not be in harmony with the spirit of this language if in some other language a word occurred which could not be translated into it; it could be claimed that 'if we can speak about anything at all, we can also speak about it in colloquial language'. If we are to maintain this universality of everyday language in connexion with semantical investigations, we must, to be

consistent, admit into the language, in addition to its sentences and other expressions, also the names of these sentences and expressions, and sentences containing these names, as well as such semantic expressions as 'true sentence' . . . .[12]

Once all this is allowed into the language, semantic antinomies result. The ideal of a theory of truth for a natural language in a natural language is therefore unattainable if we restrict ourselves to Tarski's methods. The question then arises, how to give up as little as possible, and here theories allowed by Convention T seem in important respects optimal. Such theories can bestow the required properties on a truth predicate while drawing on no conceptual resources not in the language to which the predicate applies. It is only the truth predicate itself (and the satisfaction predicate) that cannot be in the object language. Here it is essential to require no more than a *theory* of truth; to go beyond to an explicit definition does widen the gap between the resources of the object language and of the metalanguage.[13] But if we ask for no more than a theory of the kind we have been discussing, then the ontology of the metalanguage can be the same as that of the object language, and the increase in ideology can be limited to the semantical concepts.

Tarski is right, I think, in proposing that we think of natural languages as essentially intertranslatable (although I don't see why this requires word-by-word translation). The proposal idealizes the flexibility and expandability of natural languages, but can be justified by a transcendental argument (which I shall not give here).[14] Convention T demands that each sentence of the object language have a translation in the metalanguage. The pressure from the opposite direction derives from the desire to give an account, as far as possible, of the linguistic resources we command.

There are theories stated in a wholly extensional metalanguage that attempt to illuminate the semantic features of an intensional object language. The interest of such theories depends on the assumption that the object language reflects important features of a natural language. But then there is a question how we understand the metalanguage. For there is a clear sense in which the metalanguage exceeds the expressive power of the object language. For

[12] A. Tarski, 'The Concept of Truth in Formalized Languages', 164.

[13] For an instructive example, see W. V. Quine, 'On an Application of Tarski's Theory of Truth'.

[14] But see Essays 13 and 14.

each 'intensional' sentence of the object language there is a corresponding extensional sentence (about possible worlds, counterparts, etc.) with the same truth conditions in the metalanguage; but there must be metalinguistic sentences with intuitively the same subject matter that have no corresponding sentences in the object language. Intensionality is treated by such theories as a lack of expressive power—a lack we are taught how to make up for by the theory itself.[15] But if we understand our metalanguage, we are using a system of concepts and a language which is the one for which we *really* want a theory, for it is this richer system that is our natural one. And fortunately the richer system does not raise any difficulties for a truth theory satisfying Convention T, for it is extensional. (I do not consider here whether we really do understand such metalanguages.)

If a semantic theory claims to apply, however schematically, to a natural language, then it must be empirical in character, and open to test. In these concluding pages I should like to sketch my reasons for thinking that a theory that satisfies Convention T is verifiable in an interesting way.

Of course, a theory of truth is not treated as empirical if its adequacy is judged only in terms of the T-sentences it entails, and T-sentences are verified only by their form; this happens if we *assume* the object language is contained in the metalanguage. When this assumption is relaxed, the theory can go empirical. This it does the moment it is claimed that the theory applies to the speech of a particular person or group. The fact that the object language is contained in the metalanguage does not keep the theory from having empirical content: rather this fact can qualify as the fact to be verified.

In the cases that count, then, we cannot assume that described and describing languages coincide. Indeed, we cannot use a formal criterion of translation either without begging the question of empirical application. But then what becomes of Convention T? How is a T-sentence to be recognized, let alone recognized for true?

I suggest that it may be enough to require that the T-sentences be true. Clearly this suffices uniquely and correctly to determine the extension of the truth predicate. If we consider any one T-sentence, this proposal requires only that if a true sentence is described as true,

[15] I owe this insight to John Wallace.

then its truth conditions are given by some true sentence. But when we consider the constraining need to match truth with truth throughout the language, we realize that any theory acceptable by this standard may yield, in effect, a usable translation manual running from object language to metalanguage. The desired effect is standard in theory building: to extract a rich concept (here something reasonably close to translation) from thin bits of evidence (here the truth values of sentences) by imposing a formal structure on enough bits. If we characterize T-sentences by their form alone, as Tarski did, it is possible, using Tarski's methods, to define truth using no semantical concepts. If we treat T-sentences as verifiable, then a theory of truth shows how we can go from truth to something like meaning—enough like meaning so that if someone had a theory for a language verified in the way I propose, he would be able to use that language in communication.[16]

What makes it plausible that the range of theories acceptable by this standard would exclude intuitively worthless theories? One of the chief merits of the approach, that it touches the observable only when it comes to sentences, seems to leave the analysis of internal structure simply up for grabs. The articulation of sentences into singular terms, quantifiers, predicates, connectives, as well as the linking of expressions with entities in the characterization of satisfaction, must be treated as so much theoretical construction, to be tested only by its success in predicting the truth conditions of sentences.[17]

One important, indeed essential, factor in making a truth theory a credible theory of interpretation is relativization to speaker and time. When indexical or demonstrative elements are present, it cannot be sentences that are true or false, but only sentences relative to a speaker and to a time. Alternatively, we might keep truth as a property, not of sentences, but of utterances or speech acts. These in turn could be identified with certain ordered triples of sentences, times, and speakers. There are subtle problems involved in choosing between these alternatives, but I must let them go at present. The point now is that a theory that makes the right sentences true at the right times and for the right speakers will be much closer to a theory that interprets the sentences correctly than one that can ignore the

[16] For elaborations and qualifications see Essays 9–12.
[17] This point is further pursued in Essays 15 and 16.

extra parameters. In fact, something directly related to the extension of an open sentence 'Fx' will be given when we know the truth conditions of 'That is an x such that Fx'.

The metalinguistic sentence that gives the (relativized) truth conditions of a sentence with indexicals cannot have the form of a T-sentence, even if we abandon the requirement of a purely syntactical criterion for the relation between sentence described and sentence used. The trouble is that the variables ranging over persons and times, the variables introduced to accommodate the relativization, must appear in the statement of truth conditions. Truth relative to a time and a person would seem to be in the same boat as truth in a model.

It follows that the interesting difference between a theory of truth in a model and a theory of 'absolute' truth cannot be described, as I have been pretending, by saying that in the latter, but not the former, truth is characterized for each sentence *s* without appeal to any resources not in *s* itself. The reason, as we have just seen, is that 'absolute' truth goes relative when applied to a natural language. A difference remains however, and it is one to which I would attach much importance. The verification of instances of the T-sentences, or rather their surrogates in a theory relativized to speakers and times, remains respectably empirical. No doubt some pragmatic concept of *demonstration* as between speakers, times, and objects will come into play. But such a concept is one we may hope to explain without appeal to notions like truth, meaning, synonymy, or translation. The same cannot be said for truth in a model. Convention T, even when bent to fit the awkward shapes of natural language, points the way to a radical theory of interpretation.

APPLICATIONS

6 *Quotation*

Quotation is a device used to refer to typographical or phonetic shapes by exhibiting samples, that is, inscriptions or utterances that have those shapes. This characterization is broad and vague: broad enough to include not only written quotation marks, and spoken phrases like 'and I quote', but also the finger-dance quotes often used by philosophers condemned to read aloud what they have written; and vague enough to leave open the question whether the words that began this sentence ('This characterization') show a form of quotation.

In quotation not only does language turn on itself, but it does so word by word and expression by expression, and this reflexive twist is inseparable from the convenience and universal applicability of the device. Here we already have enough to draw the interest of the philosopher of language; but one discerns as well connections with further areas of concern such as sentences about propositional attitudes, explicit performatives, and picture theories of reference. If the problems raised by quotation appear trivial by comparison, we may welcome finding an easy entrance to the labyrinth.

When I was initiated into the mysteries of logic and semantics, quotation was usually introduced as a somewhat shady device, and the introduction was accompanied by a stern sermon on the sin of confusing the use and mention of expressions. The connection between quotation on the one hand and the use–mention distinction on the other is obvious, for an expression that would be used if one of its tokens appeared in a normal context is mentioned if one of its tokens appears in quotation marks (or some similar contrivance for quotation). The invitation to sin is perhaps accounted for by the ease with which quotation marks may be overlooked or omitted. But the strictures on quotation often sound a darker note. Thus Tarski, in 'The Concept of Truth in Formalized Languages', examines the possibilities for an articulate theory of quotation marks, and decides that one is led at once to absurdities, ambiguities, and contradiction.[1] Quine writes in *Mathematical Logic*, 'Scrupulous use of quotation marks is the main practical measure against confusing objects with their names . . .' but then he adds that quotation

> . . . has a certain anomalous feature which calls for special caution: from the standpoint of logical analysis each whole quotation must be regarded as a single word or sign, whose parts count for no more than serifs or syllables. A quotation is not a *description*, but a *hieroglyph*; it designates its object not only by describing it in terms of other objects, but by picturing it. The meaning of the whole does not depend upon the meanings of the constituent words.[2]

And Church, while praising Frege for his careful use of quotation to avoid equivocation, himself eschews quotation as 'misleading', 'awkward in practice . . . and open to some unfortunate abuses and misunderstandings'.[3] There is more than a hint, then, that there is something obscure or confused about quotation. But this can't be right. There is nothing wrong with the device itself. It is our theories about how it works that are inadequate or confused.

It is often said that in quotation, the quoted expressions are mentioned and not used. The first part of this claim is relatively clear. It is the second part, which says quoted expressions aren't used, that seems suspect. Why isn't incorporation into quotation one use of an expression? A plausible response would be that of course there is *some* sense in which the quoted material is used, but its use in quotation is unrelated to its *meaning* in the language; so the quoted material is not used as a piece of language.

This response may not quite still our doubts. For one thing, there are the troublesome cases where it is convenient both to use and to mention the same expression by speaking or inscribing a single token of the expression. I once resolved to adopt a consistent way of using quotation in my professional writing. My plan was to use single quotation marks when I wanted to refer to the expression a token of which was within, but double quotation marks when I wanted to use the expression in its usual meaning while at the same time indicating

[1] A. Tarski, 'The Concept of Truth in Formalized Languages', 159–62.
[2] W. V. Quine, *Mathematical Logic*, Ch. 4.
[3] A. Church, *Introduction to Mathematical Logic*, Ch. 8.

that the word was odd or special ('scare quotes'). I blush to admit that I struggled with this absurd and unworkable formula for a couple of years before it dawned on me that the second category contained the seeds of its own destruction. Consider, for example, a passage earlier in this paper where I say, nearly enough:

Quine says that quotation '. . . has a certain anomalous feature'.

Are the quoted words used or mentioned? Obviously mentioned, since the words are Quine's own, and I want to mark the fact. But equally obvious is the fact that the words are used; if they were not, what follows the word 'quotation' would be a singular term, and this is cannot be if I have produced a grammatical sentence. Nor is it easy to rephrase my words so as to resolve the difficulty. For example, it is not enough to write, 'Quine used the words "has a certain anomalous feature" of quotation', for this leaves out what he meant by those words.

Here is another mixed case of use and mention that is not altogether easy to sort out:

Dhaulagiri is adjacent to Anapurna, the mountain whose conquest Maurice Herzog described in his book of the same name.

The last phrase 'the same name' cannot mean the same name as the mountain, for the mountain has many names. Rather it means the same name of the mountain as the one used earlier in the sentence. I would call this a genuine case of quotation, for the sentence refers to an expression by exhibiting a token of that expression; but it is a case that manages without quotation marks.

Or consider this case:

The rules of Clouting and Dragoff apply, in that order.[4]

Temporarily setting aside these last examples as pathological and perhaps curable, there is a way, now standard, of giving support to the idea that in quotation the quoted material is not used. This is the interpretation of quotation proposed by Tarski as the only one he can defend. According to it a quotation, consisting of an expression flanked by quotation marks, is like a single word and is to be regarded as logically simple. The letters and spaces in the quoted material are viewed as accidents in the spelling of a longer word and

[4] The example is from J. R. Ross, 'Metalinguistic Anaphora', 273.

hence as meaningless in isolation. A quotation mark name is thus, Tarski says, like the proper name of a man.[5] I shall call this the *proper-name theory* of quotation. Church attributes the same idea, or at least a method with the same consequences, to Frege. Church writes:

> Frege introduced the device of systematically indicating autonomy by quotation marks, and in his later publications (though not in the *Begriffsschrift*) words and symbols used autonomously are enclosed in single quotation marks in all cases. This has the effect that a word enclosed in single quotation marks is to be treated as a different word from that without the quotation marks—as if the quotation marks were two additional letters in the spelling of the word—and equivocacy is thus removed by providing two different words to correspond to the different meanings.[6]

Unless I am mistaken, this passage exhibits a common confusion. For what expression is it, according to the view Church attributes to Frege, that refers to the word a token of which appears inside the quotation marks? Is it that word itself (given the context), or the quotation as a whole? Church says both, though they cannot be identical. The word itself, since an expression in quotation marks has a meaning distinct from its usual meaning; it is 'treated as a different word' which is used 'autonomously'. The quotation as a whole, since the quotation marks are part of the spelling.

Quine has repeatedly and colourfully promoted the idea of the quotation as unstructured singular term. Not only is there his denial, already cited, that quotations are descriptions, but the claim that the letters inside the quotation marks in a quotation occur there '. . . merely as a fragment of a longer name which contains, beside this fragment, the two quotation marks'.[7]

The merit in this approach to quotation is the emphasis it puts on the fact that the reference of a quotation cannot be construed as owed, at least in any normal way, to the reference of the expressions displayed within the quotation marks. But it seems to me that as an account of how quotation works in natural language, the approach is radically deficient. If quotations are structureless singular terms, then there is no more significance to the *category* of quotation-mark names than to the category of names that begin and end with the

[5] A. Tarski, 'The Concept of Truth in Formalized Languages', 159.

[6] A. Church, *Introduction to Mathematical Logic*, 61–2.

[7] Q. V. Quine, *From a Logical Point of View*, 140. Compare Quine, *Methods of Logic*, 38 and *Word and Object*, 143; also B. Mates, *Elementary Logic*, 24.

letter 'a' ('Atlanta', 'Alabama', 'Alta', 'Athena', etc.). On this view, there is no relation, beyond an accident of spelling, between an expression and the quotation-mark name of that expression. If we accept this theory, nothing would be lost if for each quotation-mark name we were to substitute some unrelated name, for that is the character of proper names. And so no echo remains, as far as this theory of quotation goes, of the informal rules governing quotation that seem so clear: if you want to form a quotation-mark name of an expression, flank that expression with quotation marks; and, a quotation-mark name refers to 'its interior' (as Quine puts it). Nothing left, either, of the intuitively attractive notion that a quotation somehow pictures what it is about.

These objections are in themselves enough to throw doubt on Tarski's claim that this interpretation of quotation is '. . . the most natural one and completely in accordance with the customary way of using quotation marks . . .'.[8] But there is a further and, I think, decisive objection, which is that on this theory we cannot give a satisfactory account of the conditions under which an arbitrary sentence containing a quotation is true. In an adequate theory, every sentence is construed as owing its truth or falsity to how it is built from a finite stock of parts by repeated application of a finite number of modes of combination. There are, of course, an infinite number of quotation-mark names, since every expression has its own quotation-mark name, and there are an infinite number of expressions. But on the theory of quotation we are considering, quotation-mark names have no significant structure. It follows that a theory of truth could not be made to cover generally sentences containing quotations. We must reject the proper-name interpretation of quotation if we want a satisfactory theory for a language containing quotations.[9]

I turn now to a quite different theory of quotation, which may be called the *picture theory* of quotation. According to this view, it is not the entire quotation, that is, expression named plus quotation marks that refers to the expression, but rather the expression itself. The role of the quotation marks is to indicate how we are to take the expression within: the quotation marks constitute a linguistic environment within which expressions do something special. This was perhaps the view of Reichenbach, who said that quotation

[8] A. Tarski, 'The Concept of Truth in Formalized Languages', 160.
[9] See Essay 1.

marks '. . . transform a sign into a name of that sign'.[10] Quine also suggests this idea when he writes that a quotation '. . . designates its object . . . by picturing it',[11] for of course it is only the interior of a quotation that could be said to be like the expression referred to (the quotation marks are not in the picture—they are the frame). And Church also, in the passage just discussed, toys with the notion that on Frege's theory '. . . a word enclosed in single quotation marks is to be treated as a different word' in that it is used 'autonomously', that is, to name itself.

It should be allowed that the three authors just mentioned, in the passages alluded to, vacillate between the proper-name theory of quotation and the picture theory. Yet the theories are clearly distinct; so bearing in mind the deficiencies of the proper-name theory, we ought to consider the picture theory on its own. At first sight it promises two advantages: it attributes *some* structure to quotations, since it treats them as composed of quotation marks (which set the scene for interpreting their contents) and the quoted material. And it hints, in its appeal to the relation of picturing, at a theory that will draw on our intuitive understanding of how quotation works.

These seeming advantages fade when examined. The difficulty is this. What is wanted is an explanation of how quotation enables us to refer to expressions by picturing them. But on the present theory, quotation marks create a context in which expressions refer to themselves. How then does picturing feature in the theory? If an expression inside quotation marks refers to itself, the fact that it also pictures itself is simply a diverting irrelevancy.

Would it help to say that quotation marks create a context in which we are to view the contents as a picture of what is referred to? Not at all; this is merely a tendentious way of saying the expression refers to itself. In brief, once the content of the quotation is assigned a standard linguistic role, the fact that it happens to resemble something has no more significance for semantics than onomatopoeia or the fact that the word 'polysyllabic' is polysyllabic.

Another important point might escape us here. The picturing relation as between an object and itself is hardly interesting and the theory, as we are interpreting it, tries vainly to make something of

[10] H. Reichenbach, *Elements of Symbolic Logic*, 335. Reichenbach says other things that point to a different theory.

[11] W. V. Quine, *Mathematical Logic*, Ch. 4.

this drab idea. But the more interesting picturing we sense in quotation is not between expression and expression. In quotation, what allows us to refer to a certain expression, which we may take to be an abstract shape, is the fact that we have before us on the page or in the air something that *has* that shape—a token, written or spoken. The picture theory suggests no way to bring an inscription or utterance into the picture. This could be done only by describing, naming, or pointing out the relevant *token*, and no machinery for the purpose has been introduced.

The picture theory of quotation is reminiscent of Frege's theory of opaque (what he called oblique) contexts such as those created by 'necessarily', 'Jones believes that . . .', 'Galileo said that . . .', and so on. There are conspicuous differences between these contexts as analysed by Frege, and quotation as treated by the picture theory: in quotation words may change their part of speech (since every expression becomes a name or description) while in the other contexts this never happens; and in quotation, but not in other opaque contexts, nonsense makes sense. But there is the striking similarity that in both cases some linguistic device is supposed to create a context within which words play new referential roles. This concept of a context that alters reference has never been properly explained, and Frege himself was leery of it: it certainly does not lend itself to direct treatment in a theory of truth. The trouble with the picture theory, as with Frege's treatment of opaque contexts generally, is that the references attributed to words or expressions in their special contexts are not functions of their references in ordinary contexts, and so the special context-creating expressions (like quotation marks or the words 'said that') cannot be viewed as functional expressions.[12]

A central defect of the proper-name theory of quotation was that while viewing quotations as well-formed expressions of the language, it failed to provide an articulate theory showing how each of the infinitude of such expressions owed its reference to its structure. The experiment just concluded showed that it is possible to treat quotations as having semantically significant structure. Let us press on in this direction.

Geach has long insisted that quotations are really *descriptions*, and hence have structure, and he complains of the proper-name

[12] For further discussion on this point, see Essay 7.

theory as I have[13] (though he does not connect his complaints with the need for a theory of truth). His theory, as I understand it, is this. A single word in quotation marks names itself; this is a new item of vocabulary, and is not semantically complex (I am not sure whether Geach says this last). So far, the theory is like the proper-name theory. But a longer expression when quoted is a structured description. Thus '"Alice swooned"' abbreviates '"Alice" ⁀ "swooned"' which reads 'the expression got by writing "'Alice'" followed by "'swooned'"'. This theory has the advantages of the preceding Fregean theory, and is far simpler and more natural. (It may be called the *spelling theory* of quotation.)

Both Tarski and Quine imply, by things they say, that they see the possibility of a similar theory. Thus Tarski remarks that if we accept the name theory, then quotation-mark names can be eliminated and replaced everywhere by structural-descriptive names,[14] while Quine contends that we can dispel the opacity of quotation, when we please, by resorting to spelling.[15] The device both have in mind is like Geach's except that Geach takes the smallest units to be words, while Tarski and Quine take them to be individual letters and symbols. The result, in the abbreviations of ordinary quotation, is the same. In primitive notation, which reveals all structure to the eye, Geach has an easier time writing (for only each word needs quotation marks) but a harder time learning or describing the language (he has a much large primitive vocabulary—twice normal size if we disregard iteration).

There is no difficulty about extending a truth definition to the devices of spelling suggested by Quine, Tarski, and Geach; yet these devices can be thought of as merely abbreviated by ordinary quotation. This claim of mere abbreviation may be backed by describing a mechanical method for going back and forth between the two styles of notation. Thus given the quotation-mark name "'Alice swooned'", the machine starts at the left by reproducing the first quotation marks, then the letter 'A', then another set of quotation marks, then a sign for concatenation, then another set of quotation marks, and so on until it reaches a set of quotations marks in the original. It reproduces these and stops. The result will be:

[13] P. T. Geach, *Mental Acts*, 79 ff. Compare Geach, 'Quotation and Quantification', 205–9.

[14] A. Tarski, 'The Concept of Truth in Formalized Languages', 160.

[15] W. V. Quine, *Word and Object*, 212.

‘A’⁀‘l’⁀‘i’⁀‘c’⁀‘e’⁀space⁀‘s’⁀‘w’⁀‘o’⁀‘o’⁀‘n’⁀‘e’⁀‘d’

Since the two notations are mechanically interchangeable, there is no reason not to consider a semantics for one a semantics for the other: so this *could* be regarded as a theory of how quotation works in English (modifications would work for other languages). But would it be a correct theory of ordinary quotation? There are several reasons for saying it would not.

Notice first that the appearance of quotation marks in the expanded notation is adventitious. The theory works by identifying a finite set of units (words or letters) from which every expression in the language to be described is composed. Then unstructured proper names of these units are introduced, along with a notation for concatenation. Such a theory works as well, and is less misleading, if quotation marks are dropped entirely and new names of the building blocks are introduced. To illustrate (following Geach's method), suppose the word ‘Alice’ is named by the word ‘alc’ and the word ‘swooned’ by the word ‘sw’; then ‘Alice swooned’ would be described by:

alc⁀sw

or, using Quine's method:

Ay⁀ell⁀cye⁀see⁀ee⁀space⁀es⁀double-you⁀oh⁀oh⁀en⁀ee⁀dee

This tiny exercise is meant to emphasize the fact that nothing of the idea of quotation *marks* is captured by this theory—nothing of the idea that one can form the name of an arbitrary expression by enclosing it in quotation marks. On the spelling theory, no articulate item in the vocabulary corresponds to quotation marks, and so the theory cannot reflect a rule for this use. The machine simply knows by heart the name of each smallest expression. Clearly, one essential element in the idea that quotations picture what they are about has been lost.

A striking way to see what is and what is not relevant to structure is to try applying existential generalization and substitution of identity. A standard way of demonstrating that quotation as normally used does not wear its structure on its surface is to observe that from:

‘Alice swooned’ is a sentence

we cannot infer:

(∃*x*)('*x* swooned' is a sentence).

Nor, supposing 'alc' names 'Alice', can we infer:

'alc swooned' is a sentence

nor:

alc⁀'swooned' is a sentence.

But (using Geach's version of the spelling theory) we *can* go from:

'Alice swooned' is a sentence

to:

'Alice'⁀'swooned' is a sentence

and thence to:

alc⁀'swooned' is a sentence

and then to:

(∃*x*)(alc⁀*x* is a sentence)

or:

(∃*x*)(*x*⁀'swooned' is a sentence).

In Quine's version of the theory, we could go from:

'Alice' is a word

to:

'A'⁀'l'⁀'i'⁀'c'⁀'e' is a word

to:

(∃*x*)(∃*y*)(*x*⁀'l'⁀'i'⁀*y*⁀'e' is a word).

These derivations show clearly that quotation marks play no vital role in the spelling theory; and also that this theory is not a theory of how quotation works in natural language.

One essential element of picturing has been lost, but not perhaps quite all, for the spelling theory does appear to depend on having the description of complex expressions reproduce the *order* of the expressions described. In the description provided by the theory, names of particular expressions need not resemble what they name, but in the description as a whole, names of expressions that are concatenated are themselves concatenated.

Even this residue of the picturing idea is superficial, however. The descriptions the spelling theory provides are themselves, from the point of view of a fully articulate language, mere abbreviations of something more complicated in which the order of expressions may well be changed. I think we should conclude that the spelling theory of quotation has no connection with the view that we understand quotations as picturing expressions.

There are further important uses of quotation in a natural language that cannot be explained by the spelling theory and could not be accommodated by a language constructed in the way it suggests. The spelling theory cannot, at least in any obvious way, deal with those mixed cases of use and mention we discussed earlier, nor indeed with any case that seems to depend on a demonstrative reference to an utterance or inscription. An important use for quotation in natural language is to introduce new notation by displaying it between quotation marks; this is impossible on the spelling theory provided the new notation is not composed of elements that have names. On the spelling theory we also could not use quotation to teach a foreign language based on a new alphabet or notation, for example Khmer or Chinese. Since these are functions easily performed by ordinary quotation (whether or not with quotation marks), we cannot accept the spelling theory as giving an adequate account of quotation in natural language.

We have discovered a short list of conditions to be satisfied by a competent theory of quotation. The first is that like a theory for any aspect of a language it should merge with a general theory of truth for the sentences of the language. The other conditions are specific to quotation. One is that the theory provide an articulate semantic role for the *devices* of quotation (quotation marks, or verbal equivalents). When we learn to understand quotation we learn a rule with endless applications: if you want to refer to an expression, you may do it by putting quotation marks around a token of the expression you want to mention. A satisfactory theory must somehow embody or explain this piece of lore. And finally, a satisfactory theory must explain the sense in which a quotation pictures what is referred to, otherwise it will be inadequate to account for important uses of quotation, for example, to introduce novel pieces of notation and new alphabets.

It is not hard to produce a satisfactory theory once the requirements are clear. The main difficulty springs, perhaps it is now obvious, from the simultaneous demands that we assign articulate structure to quotations and that they picture what they mention. For articulate linguistic structure here must be that of description, and describing seems to forestall the need to picture. The call for structure is derived from the underlying demand for a theory of meaning, here thought of as a theory of truth; all that is needed is enough structure to implement the recursive characterization of a truth predicate. Still, enough structure will be too much as long as we regard the quoted material as part of the semantically significant syntax of a sentence. The cure is therefore to give up this assumption.

It is natural to assume that words that appear between the boundaries of a sentence are legitimate parts of the sentence; and in the case of quotations, we have agreed that the words within quotation marks help us to refer to those words. Yet what I propose is that those words within quotation marks are not, from a semantical point of view, part of the sentence at all. It is in fact confusing to speak of them as words. What appears in quotation marks is an *inscription*, not a shape, and what we need it for is to help refer to its shape. On my theory, which we may call the *demonstrative theory* of quotation, the inscription inside does not refer to anything at all, nor is it part of any expression that does. Rather it is the quotation marks that do all the referring, and they help refer to a shape by pointing out something that has it. On the demonstrative theory, neither the quotation as a whole (quotes plus filling) nor the filling alone is, except by accident, a singular term. The singular term is the quotation marks, which may be read 'the expression a token of which is here'. Or, to bring out the way in which picturing may now be said genuinely to be involved: 'the expression with the shape here pictured'.

It does not discredit this theory to say that it neglects the fact that the quoted material is syntactically part of the sentence; taken in abstraction from semantics, the question of location is trivial. In spoken sentences, temporal sequence plays the role of linear arrangement in writing. But if I say 'I caught a fish this big' or 'I caught this fish today', my hands, or the fish, do not become part of the language. We could easily enough remove the quoted material from the heart of the sentence. Quotation is a device for pointing to inscriptions (or utterances) and can be used, and often is, for pointing to inscriptions or utterances spatially or temporally outside the quoting sentence. So if I follow a remark of yours with 'Truer words were never spoke', I refer to an expression, but I do it by way of indicating an embodiment of those words in an utterance. Quotation marks could be warped so as to remove the quoted material from a sentence in which they play no semantic role. Thus instead of:

> 'Alice swooned' is a sentence

we could write:

> Alice swooned. The expression of which this is a token is a sentence.

Imagine the token of 'this' supplemented with fingers pointing to the token of 'Alice swooned'.

I take it to be obvious that the demonstrative theory assigns a structure to sentences containing quotations that can be handled in a straightforward way by a theory of truth—assuming of course that there is a way of accommodating demonstratives at all, and on this point I have already tried to indicate why there is not any real difficulty in making room for demonstrative or indexical elements in a formal theory of truth.[16] Finally, it is obvious that the picturing feature of quotation has been exploited and explained. So the demonstrative theory also authorizes the use of quotation in introducing new bits of typography and discussing languages with new alphabets. I conclude by considering how it fares with the mixed cases of use and mention on exhibit earlier.

I said that for the demonstrative theory the quoted material was no part, semantically, of the quoting sentence. But this was stronger than necessary or desirable. The device of pointing can be used on whatever is in range of the pointer, and there is no reason why an inscription in active use can't be ostended in the process of mentioning an expression. I have already indicated an important sort of case, and there are many more. ('You pay attention to what I'm going to say.' 'Why did you use those words?' etc.) Any token

[16] See Essay 2.

may serve as target for the arrows of quotation, so in particular a quoting sentence may after all by chance contain a token with the shape needed for the purposes of quotation. Such tokens then do double duty, once as meaningful cogs in the machine of the sentence, once as semantically neutral objects with a useful form. Thus:

> Quine says that quotation '. . . has a certain anomalous feature'.

may be rendered more explicitly:

> Quine says, using words of which these are a token, that quotation has a certain anomalous feature.

(Here the 'these' is accompanied by a pointing to the token of Quine's words.) As for Anapurna:

> Dhaulighiri is adjacent to Anapurna, the mountain whose conquest Maurice Herzog described in his book with a name that has this shape (pointing to the token of 'Anapurna').

Finally:

> The rules of Clouting and Dragoff apply, in the order in which these tokens appear (pointing to the tokens of 'Clouting' and 'Dragoff').

7 *On Saying That*

'I wish I had said that', said Oscar Wilde in applauding one of Whistler's witticisms. Whistler, who took a dim view of Wilde's originality, retorted, 'You will, Oscar; you will'.[1] This tale reminds us that an expression like 'Whistler said that' may on occasion serve as a grammatically complete sentence. Here we have, I suggest, the key to a correct analysis of indirect discourse, an analysis that opens a lead to an analysis of psychological sentences generally (sentences about propositional attitudes, so-called), and even, though this looks beyond anything to be discussed in the present paper, a clue to what distinguishes psychological concepts from others.

But let us begin with sentences usually deemed more representative of *oratio obliqua*, for example 'Galileo said that the earth moves' or 'Scott said that Venus is an inferior planet'. One trouble with such sentences is that we do not know their logical form. And to admit this is to admit that, whatever else we may know about them, we do not know the first thing. If we accept surface grammar as guide to logical form, we will see 'Galileo said that the earth moves' as containing the sentence 'the earth moves', and this sentence in turn as consisting of the singular term 'the earth', and a predicate, 'moves'. But if 'the earth' is, in this context, a singular term, it can be replaced, so far as the truth or falsity of the containing sentence is concerned, by any other singular term that refers to the same thing. Yet what seem like appropriate replacements can alter the truth of the original sentence.

The notorious apparent invalidity of this move can only be apparent, for the rule on which it is based no more than spells out

[1] From H. Jackson, *The Eighteen-Nineties*, 73.

what is involved in the idea of a (logically) singular term. Only two lines of explanation, then, are open: we are wrong about the logical form, or we are wrong about the reference of the singular term.

What seems anomalous behaviour on the part of what seem singular terms dramatizes the problem of giving an orderly account of indirect discourse, but the problem is more pervasive. For what touches singular terms touches what they touch, and that is everything: quantifiers, variables, predicates, connectives. Singular terms refer, or pretend to refer, to the entities over which the variables of quantification range, and it is these entities of which the predicates are or are not true. So it should not surprise us that if we can make trouble for the sentence 'Scott said that Venus is an inferior planet' by substituting 'the Evening Star' for 'Venus', we can equally make trouble by substituting 'is identical with Venus or with Mercury' for the coextensive 'is an inferior planet'. The difficulties with indirect discourse cannot be solved simply by abolishing singular terms.

What should we ask of an adequate account of the logical form of a sentence? Above all, I would say, such an account must lead us to see the semantic character of the sentence—its truth or falsity—as owed to how it is composed, by a finite number of applications of some of a finite number of devices that suffice for the language as a whole, out of elements drawn from a finite stock (the vocabulary) that suffices for the language as a whole. To see a sentence in this light is to see it in the light of a theory for its language, a theory that gives the form of every sentence in that language. A way to provide such a theory is by recursively characterizing a truth predicate, along the lines suggested by Tarski.[2]

Two closely linked considerations support the idea that the structure with which a sentence is endowed by a theory of truth in Tarski's style deserves to be called the logical form of the sentence. By giving such a theory, we demonstrate in a persuasive way that the language, though it consists in an indefinitely large number of sentences, can be comprehended by a creature with finite powers. A theory of truth may be said to supply an effective explanation of the semantic role of each significant expression in any of its appearances. Armed with the theory, we can always answer the question, 'What are these familiar words doing here?' by saying how they

[2] A. Tarski, 'The Concept of Truth in Formalized Languages'. See Essays 2, 4, and 5.

contribute to the truth conditions of the sentence. (This is not to assign a 'meaning', much less a reference, to every significant expression.)

The study of the logical form of sentences is often seen in the light of another interest, that of expediting inference. From this point of view, to give the logical form of a sentence is to catalogue the features relevant to its place on the logical scene, the features that determine what sentences it is a logical consequence of, and what sentences it has as logical consequences. A canonical notation graphically encodes the relevant information, making theory of inference simple, and practice mechanical where possible.

Obviously the two approaches to logical form cannot yield wholly independent results, for logical consequence is defined in terms of truth. To say a second sentence is a logical consequence of a first is to say, roughly, that the second is true if the first is no matter how the non-logical constants are interpreted. Since what we count as a logical constant can vary independently of the set of truths, it is clear that the two versions of logical form, though related, need not be identical. The relation, in brief, seems this. Any theory of truth that satisfies Tarski's criteria must take account of all truth-affecting iterative devices in the language. In the familiar languages for which we know how to define truth the basic iterative devices are reducible to the sentential connectives, the apparatus of quantification, and the description operator if it is primitive. Where one sentence is a logical consequence of another on the basis of quantificational structure alone, a theory of truth will therefore entail that if the first sentence is true, the second is. There is no point, then, in not including the expressions that determine quantificational structure among the logical constants, for when we have characterized truth, on which any account of logical consequence depends, we have already committed ourselves to all that calling such expressions logical constants could commit us. Adding to this list of logical constants will increase the inventory of logical truths and consequence-relations beyond anything a truth definition demands, and will therefore yield richer versions of logical form. For the purposes of the present paper, however, we can cleave to the most austere interpretations of logical consequence and logical form, those that are forced on us when we give a theory of truth.[3]

[3] For further defence of a concept of logical form based on a theory of truth, see *Essays on Actions and Events*, 137–46.

We are now in a position to explain our aporia over indirect discourse: what happens is that the relation between truth and consequence just sketched appears to break down. In a sentence like 'Galileo said that the earth moves' the eye and mind perceive familiar structure in the words 'the earth moves'. And structure there must be if we are to have a theory of truth at all, for an infinite number of sentences (all sentences in the indicative, apart from some trouble over tense) yield sense when plugged into the slot in 'Galileo said that ________'. So if we are to give conditions of truth for all the sentences so generated, we cannot do it sentence by sentence, but only by discovering an articulate structure that permits us to treat each sentence as composed of a finite number of devices that make a stated contribution to its truth conditions. As soon as we assign familiar structure, however, we must allow the consequences of that assignment to flow, and these, as we know, are in the case of indirect discourse consequences we refuse to buy. In a way, the matter is even stranger than that. Not only do familiar consequences fail to flow from what looks to be familiar structure, but our common sense of language feels little assurance in any inferences based on the words that follow the 'said that' of indirect discourse (there are exceptions).

So the paradox is this: on the one hand, intuition suggests, and theory demands, that we discover semantically significant structure in the 'content-sentences' of indirect discourse (as I shall call sentences following 'said that'). On the other hand, the failure of consequence-relations invites us to treat contained sentences as semantically inert. Yet logical form and consequence relations cannot be divorced in this way.

One proposal at this point is to view the words that succeed the 'said that' as operating within concealed quotation marks, their sole function being to help refer to a sentence, and their semantic inertness explained by an account of quotation. One drawback of this proposal is that no usual account of quotation is acceptable, even by the minimal standards we have set for an account of logical form. For according to most stories, quotations are singular terms without significant semantic structure, and since there must be an infinite number of different quotations, no language that contains them can have a recursively defined truth predicate. This may be taken to show that the received accounts of quotation must be mistaken—I think it does. But then we can hardly pretend that we have solved the problem of indirect discourse by appeal to quotation.[4]

Perhaps it is not hard to invent a theory of quotation that will serve: the following theory is all but explicit in Quine. Simply view quotations as abbreviations for what you get if you follow these instructions: to the right of the first letter that has opening quotation marks on its left write right-hand quotation marks, then the sign for concatenation, and then left-hand quotation marks, in that order; do this after each letter (treating punctuation signs as letters) until you reach the terminating right-hand quotation marks. What you now have is a complex singular term that gives what Tarski calls a structural description of an expression. There is a modest addition to vocabulary: names of letters and of punctuation signs, and the sign for concatenation. There is a corresponding addition to ontology: letters and punctuation signs. And finally, if we carry out the application to sentences in indirect discourse, there will be the logical consequences that the new structure dictates. For two examples, each of the following will be entailed by 'Galileo said that the earth moves':

$(\exists x)$ (Galileo said that 'the ea'$^\frown x^\frown$'th moves')

and (with the premise 'r = the 18th letter in the alphabet'):

Galileo said that 'the ea'$^\frown$the 18th letter of the alphabet$^\frown$'th moves'

(I have clung to abbreviations as far as possible.) These inferences are not meant in themselves as criticism of the theory of quotation; they merely illuminate it.

Quine discusses the quotational approach to indirect discourse in *Word and Object*,[5] and abandons it for what seems, to me, a wrong reason. Not that there is not a good reason; but to appreciate *it* is to be next door to a solution, as I shall try to show.

Let us follow Quine through the steps that lead him to reject the quotational approach. The version of the theory he considers is not the one once proposed by Carnap to the effect that 'said that' is a two-place predicate true of ordered pairs of people and sentences.[6]

[4] See Essays 1 and 6.

[5] W. V. Quine, *Word and Object*, Ch. 6. Hereafter numerals in parentheses refer to pages of this book.

[6] R. Carnap, *The Logical Syntax of Language*, 248. The same was in effect proposed by P. T. Geach in *Mental Acts*.

The trouble with this idea is not that it forces us to assimilate indirect discourse to direct, for it does not. The 'said that' of indirect discourse, like the 'said' of direct, may relate persons and sentences, but be a different relation; the former, unlike the latter, may be true of a person, and a sentence he never spoke in a language he never knew. The trouble lies rather in the chance that the same sentence may have different meanings in different languages—not too long a chance either if we count ideolects as languages. To give an example, the sounds 'Empedokles liebt' do fairly well as a German or an English sentence, in one case saying that Empedokles loved and in the other telling us what he did from the top of Etna. If we analyse 'Galileo said that the earth moves' as asserting a relation between Galileo and the sentence 'The earth moves', we do not have to assume that Galileo spoke English, but we cannot avoid the assumption that the words of the content-sentence are to be understood as an English sentence.[7]

Calling the relativity to English an assumption may be misleading; perhaps the reference to English is explicit, as follows. A long-winded version of our favourite sentence might be 'Galileo spoke a sentence that meant in his language what "The earth moves" means in English'. Since in this version it needs all the words except 'Galileo' and 'The earth moves' to do the work of 'said that', we must count the reference to English as explicit in the 'said that'. To see how odd this is, however, it is only necessary to reflect that the English words 'said that', with their built-in reference to English, would no longer translate (by even the roughest extensional standards) the French 'dit que'.

We can shift the difficulty over translation away from the 'said that' or 'dit que' by taking these expressions as three-place predicates relating a speaker, a sentence, and a language, the reference to a language to be supplied either by our (in practice nearly infallible) knowledge of the language to which the quoted material is to be taken as belonging, or by a demonstrative reference to the language of the entire sentence. Each of these suggestions has its own appeal, but neither leads to an analysis that will pass the translation test. To take the demonstrative proposal, translation into French will carry 'said that' into 'dit que', the demonstrative reference will automatically, and hence perhaps still within the

[7] The point is due to A. Church, 'On Carnap's Analysis of Statements of Assertion and Belief'.

bounds of strict translation, shift from English to French. But when we translate the final singular term, which names an English sentence, we produce a palpably false result.

These exercises help bring out important features of the quotational approach. But now it is time to remark that there would be an anomaly in a position, like the one under consideration, that abjured reference to propositions in favour of reference to languages. For languages (as Quine remarks in a similar context in *Word and Object*) are at least as badly individuated, and for much the same reasons, as propositions. Indeed, an obvious proposal linking them is this: languages are identical when identical sentences express identical propositions. We see, then, that quotational theories of indirect discourse, those we have discussed anyway, cannot claim an advantage over theories that frankly introduce intensional entities from the start; so let us briefly consider theories of the latter sort.

It might be thought, and perhaps often is, that if we are willing to welcome intensional entities without stint—properties, propositions, individual concepts, and whatever else—then no further difficulties stand in the way of giving an account of the logical form of sentences in *oratio obliqua*. This is not so. Neither the languages Frege suggests as models for natural languages nor the languages described by Church are amenable to theory in the sense of a truth definition meeting Tarski's standards.[8] What stands in the way in Frege's case is that every referring expression has an infinite number of entities it may refer to, depending on the context, and there is no rule that gives the reference in more complex contexts on the basis of the reference in simpler ones. In Church's languages, there is an infinite number of primitive expressions; this directly blocks the possibility of recursively characterizing a truth predicate satisfying Tarski's requirements.

Things might be patched up by following a leading idea of Carnap's *Meaning and Necessity* and limiting the semantic levels to two: extensions and (first-level) intensions.[9] An attractive strategy might then be to turn Frege, thus simplified, upside down by letting each singular term refer to its sense or intension, and providing a

[8] G. Frege, 'On Sense and Reference'; A. Church, 'A Formulation of the Logic of Sense and Denotation'. See Essay 1.

[9] The idea of an essentially Fregean approach limited to two semantic levels has been suggested by M. Dummett in *Frege: Philosophy of Language*, Ch. 9.

reality function (similar to Church's delta function) to map intensions on to extensions. Under such treatment our sample sentence would emerge like this: 'The reality of Galileo said that the earth moves.' Here we must suppose that 'the earth' names an individual concept which the function referred to by 'moves' maps on to the proposition that the earth moves; the function referred to by 'said that' in turn maps Galileo and the proposition that the earth moves on to a truth value. Finally, the name 'Galileo' refers to an individual concept which is mapped, by the function referred to by 'the reality of' on to Galileo. With ingenuity, this theory can perhaps be made to accommodate quantifiers that bind variables both inside and outside contexts created by verbs like 'said' and 'believes'. There is no special problem about defining truth for such a language: everything is on the up and up, purely extensional save in ontology. This seems to be a theory that might do all we have asked. Apart from nominalistic qualms, why not accept it?

My reasons against this course are essentially Quine's. Finding right words of my own to communicate another's saying is a problem in translation (216–17). The words I use in the particular case may be viewed as products of my total theory (however vague and subject to correction) of what the originating speaker means by anything he says: such a theory is indistinguishable from a characterization of a truth predicate, with his language as object language and mine as metalanguage. The crucial point is that there will be equally acceptable alternative theories which differ in assigning clearly non-synonymous sentences of mine as translations of his same utterance. This is Quine's thesis of the indeterminacy of translation (218–21).[10] An example will help bring out the fact that the thesis applies not only to translation between speakers of conspicuously different languages, but also to cases nearer home.

Let someone say (and now discourse is direct), 'There's a hippopotamus in the refrigerator'; am I necessarily right in reporting him as having said that there is a hippopotamus in the refrigerator? Perhaps; but under questioning he goes on, 'It's roundish, has a wrinkled skin, does not mind being touched. It has a pleasant taste, at least the juice, and it costs a dime. I squeeze two or three for breakfast.' After some finite amount of such talk we slip over the line where it is plausible or even possible to say correctly

[10] My assimilation of a translation manual to a theory of truth is not in Quine. For more on this and related matters, see Essays 2, 11, and 16.

that he said there was a hippopotamus in the refrigerator, for it becomes clear he means something else by at least some of his words than I do. The simplest hypothesis so far is that my word 'hippopotamus' no longer translates his word 'hippopotamus'; my word 'orange' might do better. But in any case, long before we reach the point where homophonic translation must be abandoned, charity invites departures. Hesitation over whether to translate a saying of another by one or another of various non-synonymous sentences of mine does not necessarily reflect a lack of information: it is just that beyond a point there is no deciding, even in principle, between the view that the Other has used words as we do but has more or less weird beliefs, and the view that we have translated him wrong. Torn between the need to make sense of a speaker's words and the need to make sense of the pattern of his beliefs, the best we can do is choose a theory of translation that maximizes agreement. Surely there is no future in supposing that in earnestly uttering the words 'There's a hippopotamus in the refrigerator' the Other has disagreed with us about what can be in the refrigerator if we also must then find ourselves disagreeing with him about the size, shape, colour, manufacturer, horsepower, and wheelbase of hippopotami.

None of this shows there is no such thing as correctly reporting, through indirect discourse, what another has said. All that the indeterminacy shows is that if there is one way of getting it right there are other ways that differ substantially in that non-synonymous sentences are used after 'said that'. And this is enough to justify our feeling that there is something bogus about the sharpness questions of meaning must in principle have if meanings are entities.

The lesson was implicit in a discussion started some years ago by Benson Mates. Mates claimed that the sentence 'Nobody doubts that whoever believes that the seventh consulate of Marius lasted less than a fortnight believes that the seventh consulate of Marius lasted less than a fortnight' is true and yet might well become false if the last word were replaced by the (supposed synonymous) words 'period of fourteen days', and that this could happen no matter what standards of synonomy we adopt short of the question-begging 'substitutable everywhere *salva veritate*'.[11] Church and Sellars responded by saying the difficulty could be resolved by firmly distinguishing between substitutions based on the speaker's use of

[11] B. Mates, 'Synonymity'. The example is Church's.

language and substitutions coloured by the use attributed to others.[12] But this is a solution only if we think there is some way of telling, in what another says, what is owed to the meanings he gives his words and what to his beliefs about the world. According to Quine, this is a distinction that cannot be drawn.

The detour has been lengthy; I return now to Quine's discussion of the quotational approach in *Word and Object*. As reported above, Quine rejects relativization to a language on the grounds that the principle of the individuation of languages is obscure, and the issue when languages are identical irrelevant to indirect discourse (214). He now suggests that instead of interpreting the content-sentence of indirect discourse as occurring in a language, we interpret it as voiced by a speaker at a time. The speaker and time relative to which the content-sentence needs understanding is, of course, the speaker of that sentence, who is thereby indirectly attributing a saying to another. So now 'Galileo said that the earth moves' comes to mean something like 'Galileo spoke a sentence that in his mouth meant what "The earth moves" now means in mine'. Quine makes no objection to this proposal because he thinks he has something simpler and at least as good in reserve. But in my opinion the present proposal deserves more serious consideration, for I think it is nearly right, while Quine's preferred alternatives are seriously defective.

The first of these alternatives is Scheffler's inscriptional theory.[13] Scheffler suggests that sentences in indirect discourse relate a speaker and an utterance: the role of the content-sentence is to help convey what sort of utterance it was. What we get this way is, 'Galileo spoke a that-the-earth-moves utterance'. The predicate 'x is-a-that-the-earth-moves-utterance' has, so far as theory of truth and of inference are concerned, the form of an unstructured one-place predicate. Quine does not put the matter quite this way, and he may resist my appropriation of the terms 'logical form' and 'structure' for purposes that exclude application to Scheffler's predicate. Quine calls the predicate 'compound' and describes it as composed of an operator and a sentence (214, 215). These are matters of terminology; the substance, about which there may be no disagreement, is that on Scheffler's theory sentences in *oratio obliqua* have no logical relations that depend on structure in the predicate,

[12] A. Church, 'Intensional Isomorphism and Identity of Belief'; W. Sellars, 'Putnam on Synonymity and Belief'.

[13] I. Scheffler, 'An Inscriptional Approach to Indirect Quotation'.

and a truth predicate that applies to all such sentences cannot be characterized in Tarski's style. The reason is plain: there is an infinite number of predicates with the syntax '*x* is-a-________-utterance' each of which is, in the eyes of semantic theory, unrelated to the rest.

Quine has seized one horn of the dilemma. Since attributing semantic structure to content-sentences in indirect discourse apparently forces us to endorse logical relations we do not want, Quine gives up the structure. The result is that another desideratum of theory is neglected, that truth be defined.

Consistent with his policy of renouncing structure that supports no inferences worth their keep, Quine contemplates one further step; he says, '. . . a final alternative that I find as appealing as any is simply to dispense with the objects of the propositional attitudes' (216). Where Scheffler still saw 'said that' as a two-place predicate relating speakers and utterances, though welding content-sentences into one-piece one-place predicates true of utterances, Quine now envisions content-sentence and 'said that' welded directly to form the one-place predicate '*x* said-that-the-earth-moves', true of persons. Of course some inferences inherent in Scheffler's scheme now fall away: we can no longer infer 'Galileo said something' from our sample sentence, nor can we infer from it and 'Someone denied that the earth moves' the sentence 'Someone denied what Galileo said'. Yet as Quine reminds us, inferences like these may fail on Scheffler's analysis too when the analysis is extended along the obvious line to belief and other propositional attitudes, since needed utterances may fail to materialize (215). The advantages of Scheffler's theory over Quine's 'final alternative' are therefore few and uncertain; this is why Quine concludes that the view that invites the fewest inferences is 'as appealing as any'.

This way of eliminating unwanted inferences unfortunately abolishes most of the structure needed by the theory of truth. So it is worth returning for another look at the earlier proposal to analyse indirect discourse in terms of a predicate relating an originating speaker, a sentence, and the present speaker of the sentence in indirect discourse. For that proposal did not cut off any of the simple entailments we have been discussing, and it alone of recent suggestions promised, when coupled with a workable theory of quotation, to yield to standard semantic methods. But there is a subtle flaw.

We tried to bring out the flavour of the analysis to which we have returned by rewording our favourite sentence as 'Galileo uttered a sentence that meant in his mouth what "The earth moves" means now in mine'. We should not think ill of this verbose version of 'Galileo said that the earth moves' because of apparent reference to a meaning ('what "The earth moves" means'); this expression is not treated as a singular term in the theory. We are indeed asked to make sense of a judgement of synonomy between utterances, but not as the foundation of a theory of language, merely as an unanalysed part of the content of the familiar idiom of indirect discourse. The idea that underlies our awkward paraphrase is that of *samesaying*: when I say that Galileo said that the earth moves, I represent Galileo and myself as samesayers.[14]

And now the flaw is this. If I merely *say* we are samesayers, Galileo and I, I have yet to *make* us so; and how am I to do this? Obviously, by saying what he said; not by using his words (necessarily), but by using words the same in import here and now as his then and there. Yet this is just what, on the theory, I cannot do. For the theory brings the content-sentence into the act sealed in quotation marks, and on any standard theory of quotation, this means the content-sentence is mentioned and not used. In uttering the words 'The earth moves' I do not, according to this account, say anything remotely like what Galileo is claimed to have said; I do not, in fact, say anything. My words in the frame provided by 'Galileo said that________' merely help refer to a sentence. There will be no missing the point if we expand quotation in the style we recently considered. Any intimation that Galileo and I are samesayers vanishes in this version:

Galileo said that 'T'⁀'h'⁀'e'⁀' '⁀'e'⁀'a'⁀'r'⁀'t'⁀'h'⁀' '⁀'m'⁀'o'⁀'v'⁀'e'⁀'s'

[14] Strictly speaking, the verb 'said' is here analysed as a three-place predicate which holds of a speaker (Galileo), an utterance of the speaker ('Eppur si muove'), and an utterance of the attributer ('The earth moves'). This predicate is from a semantic point of view a primitive. The fact that an informal paraphrase of the predicate appeals to a relation of sameness of content as between utterances introduces no intentional entities or semantics. Some have regarded this as a form of cheating, but the policy is deliberate and principled. For a discussion of the distinction between questions of logical form (which is the present concern) and the analysis of individual predicates, see Essay 2. It is also worth observing that radical interpretation, if it succeeds, yields an adequate concept of synonymy as between utterances. See the end of Essay 12. [Footnote added in 1982.]

We seem to have been taken in by a notational accident, a way of referring to expressions that when abbreviated produces framed pictures of the very words referred to. The difficulty is odd; let's see if we can circumvent it. Imagine an altered case. Galileo utters his words 'Eppur si muove', I utter my words, 'The earth moves'. There is no problem yet in recognizing that we are samesayers; an utterance of mine matches an utterance of his in purport. I am not now using my words to help refer to a sentence; I speak for myself, and my words refer in their usual way to the earth and to its movement. If Galileo's utterance 'Eppur si muove' made us samesayers, then some utterance or other of Galileo's made us samesayers. The form '($\exists x$) (Galileo's utterance x and my utterance y makes us samesayers)' is thus a way of attributing any saying I please to Galileo provided I find a way of replacing 'y' by a word or phrase that refers to an appropriate utterance of mine. And surely there is a way I can do this: I need only produce the required utterance and replace 'y' by a reference to it. Here goes:

The earth moves.

($\exists x$) (Galileo's utterance x and my last utterance makes us samesayers).

Definitional abbreviation is all that is needed to bring this little skit down to:

The earth moves.
Galileo said that.

Here the 'that' is a demonstrative singular term referring to an utterance (not a sentence).

This form has a small drawback in that it leaves the hearer up in the air about the purpose served by saying 'The earth moves' until the act has been performed. As if, say, I were first to tell a story and then add, 'That's how it was once upon a time'. There's some fun to be had this way, and in any case no amount of telling what the illocutionary force of our utterances is is going to insure that they have that force. But in the present case nothing stands in the way of reversing the order of things, thus:

Galileo said that.
The earth moves.

It is now safe to allow a tiny orthographic change, a change without semantic significance, but suggesting to the eye the relation of introducer and introduced: we may suppress the stop after 'that' and the consequent capitalization:

Galileo said that the earth moves.

Perhaps it should come as no surprise to learn that the form of psychological sentences in English apparently evolved in much the way these ruminations suggest. According to the *Oxford English Dictionary*,

> The use of *that* is generally held to have arisen out of the demonstrative pronoun pointing to the clause which it introduces. Cf. (1) He once lived here: we all know *that*; (2) *That* (now *this*) we all know: he once lived here; (3) We all know *that* (or *this*): he once lived here; (4) We all know *that* he once lived here . . .[15]

The proposal then is this: sentences in indirect discourse, as it happens, wear their logical form on their sleeves (except for one small point). They consist of an expression referring to a speaker, the two-place predicate 'said', and a demonstrative referring to an utterance. Period. What follows gives the content of the subject's saying, but has no logical or semantic connection with the original attribution of a saying. This last point is no doubt the novel one, and upon it everything depends: from a semantic point of view the content-sentence in indirect discourse is not contained in the sentence whose truth counts, i.e. the sentence that ends with 'that'.

We would do better, in coping with this subject, to talk of inscriptions and utterances and speech acts, and avoid reference to sentences.[16] For what an utterance of 'Galileo said that' does is announce a further utterance. Like any utterance, this first may be serious or silly, assertive or playful; but if it is true, it must be followed by an utterance synonymous with some other. The second utterance, the introduced act, may also be true or false, done in the mode of assertion or of play. But if it is as announced, it must serve

[15] J. A. H. Murray *et al.* (eds.), *The Oxford English Dictionary*, 253. Cf. C. T. Onions, *An Advanced English Syntax*, 154–6. I first learned that 'that' in such contexts evolved from an explicit demonstrative in J. Hintikka, *Knowledge and Belief*, 13. Hintikka remarks that a similar development has taken place in German and Finnish. I owe the *OED* reference to Eric Stiezel.

[16] I assume that a theory of truth for a language containing demonstratives must apply strictly to utterances and not to sentences, or will treat truth as a relation between sentences, speakers, and times. See Essays 2 and 4.

at least the purpose of conveying the content of what someone said. The role of the introducing utterance is not unfamiliar: we do the same with words like 'This is a joke', 'This is an order', 'He commanded that', 'Now hear this'. Such expressions might be called performatives, for they are used to usher in performances on the part of the speaker. A certain interesting reflexive effect sets in when performatives occur in the first-person present tense, for then the speaker utters words which if true are made so exclusively by the content and mode of the performance that follows, and the mode of this performance may well be in part determined by that same performative introduction. Here is an example that will also provide the occasion for a final comment on indirect discourse.

'Jones asserted that Entebbe is equatorial' would, if we parallel the analysis of indirect discourse, come to mean something like, 'An utterance of Jones' in the assertive mode had the content of this utterance of mine. Entebbe is equatorial.' The analysis does not founder because the modes of utterance of the two speakers may differ; all that the truth of the performative requires is that the second utterance, in whatever mode (assertive or not), match in content an assertive utterance of Jones. Whether such an asymmetry is appropriate in indirect discourse depends on how much of assertion we read into the concept of saying. Now suppose I try: 'I assert that Entebbe is equatorial.' Of course by saying this I may not assert anything; mood of words cannot guarantee mode of utterance. But if my utterance of the performative is true, then do I say something in the assertive mode that has the content of my second utterance—I do, that is, assert that Entebbe is equatorial. If I do assert it, an element in my success is no doubt my utterance of the performative, which announces an assertion; thus performatives tend to be self-fulfilling. Perhaps it is this feature of performatives that has misled some philosophers into thinking that performatives, or their utterances, are neither true nor false.

On the analysis of indirect discourse here proposed, standard problems seem to find a just solution. The appearance of failure of the laws of extensional substitution is explained as due to our mistaking what are really two sentences for one: we make substitutions in one sentence, but it is the other (the utterance of) which changes in truth. Since an utterance of 'Galileo said that' and any utterance following it are semantically independent, there is no reason to predict, on grounds of form alone, any *particular* effect on the truth of the first from change in the second. On the other hand, if the second utterance had been different in any way at all, the first utterance *might* have had a different truth value, for the reference of the 'that' would have changed.

The paradox, that sentences (utterances) in *oratio obliqua* do not have the logical consequences they should if truth is to be defined, is resolved. What follows the verb 'said' has only the structure of a singular term, usually the demonstrative 'that'. Assuming the 'that' refers, we can infer that Galileo said something from 'Galileo said that'; but this is welcome. The familiar words coming in the train of the performative of indirect discourse do, on my account, have structure, but it is familiar structure and poses no problem for theory of truth not there before indirect discourse was the theme.

Since Frege, philosophers have become hardened to the idea that content-sentences in talk about propositional attitudes may strangely refer to such entities as intensions, propositions, sentences, utterances, and inscriptions. What is strange is not the entities, which are all right in their place (if they have one), but the notion that ordinary words for planets, people, tables, and hippopotami in indirect discourse may give up these pedestrain references for the exotica. If we could recover our pre-Fregean semantic innocence, I think it would seem to us plainly incredible that the words 'The earth moves', uttered after the words 'Galileo said that', mean anything different, or refer to anything else, than is their wont when they come in other environments. No doubt their role in *oratio obliqua* is in some sense special; but that is another story. Language is the instrument it is because the same expression, with semantic features (meaning) unchanged, can serve countless purposes. I have tried to show how our understanding of indirect discourse does not strain this basic insight.

8 *Moods and Performances*

Frege held that an adequate account of language requires us to attend to three features of sentences: reference, sense, and force. Elsewhere I have argued that a theory of truth patterned after a Tarski-type truth definition tells us all we need to know about sense.[1] Counting truth in the domain of reference, as Frege did, the study of sense thus comes down to the study of reference.

But how about force? In this paper I want to consider force in the only form in which I am certain that it is a feature of sentences, that is, as it serves to distinguish the moods. The question I am concerned with is, can a theory of truth explain the differences among the moods?

In trying to answer this question I am responding, belatedly,[2] alas, to a challenge put to me by Yehoshua Bar-Hillel some years ago; he asked me how it might be possible to represent mood within the confines of a theory of truth.

One reason the analysis of mood is interesting is that it forces us to attend to the relations between what sentences mean, and their uses. We have on the one hand the syntactic, and presumably semantic, distinction among moods (such as: indicative, imperative, optative, interrogative), and on the other hand the distinction among uses of sentences (such as: to make assertions, to give orders, to express wishes, to ask questions).

The moods classify sentences, while uses classify utterances; but the moods indirectly classify utterances, since whatever distinguishes

[1] See Essays 2, 4, and 9–12 for support for, and doubts concerning, this claim.

[2] The conference at which this paper was read was to have celebrated Bar-Hillel's sixtieth birthday. A great loss to philosophy and to his many friends, he died before the conference was held.

sentences can be used to distinguish utterances of them. So we may ask, what is the relation between these two ways of classifying utterances; how are assertions related to utterances of indicative sentences, for example, or commands to utterances of imperative sentences?

The simplest suggestion would be that the associated classes of utterances are identical: utterances of imperatives are commands, utterances of interrogatives are question-askings, etc. This idea appears to find support in some of Dummett's work. Here is how Dummett explains Frege's use of the assertion sign or judgement-stroke:

> The judgment-stroke is the sign of assertion proper, that which carries the assertive force. It is therefore not a functional expression, or part of one: we cannot enquire of it what its sense is, or what its reference is; it contributes to the meaning of the complex sentential symbol in quite a different way . . . it is only the sentence to which the judgment-stroke is prefixed which may be said to express a sense or to stand for a truth-value: the whole expression with the judgment-stroke neither expresses anything nor stands for anything—it *asserts* something: it asserts, namely, that the thought expressed by what follows the judgment-stroke is true.[3]

Here Dummett says that it is sentences that make assertions, where I think it would be more natural to say that an assertion is an utterance, and it is the speaker who makes the assertion. However, this may be no more than a terminological complaint; what bothers me is the implied claim that assertion and the indicative mood can be this closely identified. For there are many utterances of indicative sentences that are not assertions, for example indicative sentences uttered in play, pretence, joke, and fiction; and of course assertions may be made by uttering sentences in other moods. (Utterances of 'Did you notice that Joan is wearing her purple hat again?' or 'Notice that Joan is wearing her purple hat again' may on occasion simply be assertions that Joan is wearing her purple hat again.) And similarly for the other moods; we can ask a question with an imperative or indicative ('Tell me who won the third race', 'I'd like to know your telephone number'), or issue a command with an indicative ('In this house we remove our shoes before entering').

Needless to say, Dummett knows all this, and if he temporarily allows himself to overlook these cases for the sake of the larger view, it is only because he believes that there is a clear sense in which the

[3] M. Dummett, *Frege: Philosophy of Language*, 315, 316.

counter-examples are deviant. But what is this sense? Austin made a distinction between what he called the 'normal' or 'serious' uses of a sentence and the 'etiolated' or 'parasitical' uses.[4] If such a distinction could be made in a non-circular way, and it turned out that the normal or serious use of indicatives was to make assertions, of imperatives to issue commands, of interrogatives to ask questions, and so on, then the desired connection between the moods and uses of sentences would be established.

There surely is some important connection between the moods and their uses, and so we are bound to think that there is something natural, serious, or normal, about using a sentence in a certain mood to perform a 'corresponding' act. The question is whether this feeling can be articulated in a way that throws light on the nature of the moods. It is easy to see that appeal to what is 'serious' or 'normal' does not go beyond an appeal to intuition. It is no clue to the seriousness of a command that it is uttered in the imperative rather than the indicative; similarly, a serious question may be posed in the imperative rather than the interrogative mood. And if 'normal' means usual, or statistically more frequent, it is dubious indeed that most indicatives are uttered as assertions. There are too many stories, rote repetitions, illustrations, suppositions, parodies, charades, chants, and conspicuously unmeant compliments. And in any case the analysis of mood cannot plausibly rest on the results of this sort of statistical survey.

Dummett's solution is to switch from the serious or normal to the conventional: an assertion is an indicative uttered under conditions specified by convention; a command is an imperative uttered under other conventionally given conditions; and so forth. So he writes, '... assertion consists in the (deliberate) utterance of a sentence which, by its form and context, is recognized as being used according to a certain general convention ...'[5] And of imperatives, '... the utterance of a sentence of a certain form, unless special circumstances divest this act of its usual significance, in itself constitutes the giving of a command'.[6] He sums up with this advice on how to approach the subject of the relations between the moods and their uses:

> ... the correct approach is to consider utterances as conventionally demarcated into types, by means of the form of linguistic expressions

[4] J. L. Austin, *How to Do Things with Words*, 22.
[5] M. Dummett, *Frege: Philosophy of Language*, 311.
[6] Ibid., 301, 302.

employed, and then to enquire into the conventions governing the use of the various types of utterance.[7]

Dummett's view that linguistic actions like assertion and command consist in uttering sentences in the indicative or imperative moods under conventionally specified conditions is central to his picture of language when coupled with the thesis that there is a further convention that assertions are made with the intention of saying what is true. For these two ideas together would establish a direct connection between languages as used in conventional ways and a certain overall purpose (to say what is true).[8]

I agree that we must find connections between how sentences are used and what they mean if we are to give a foundational account of language. I am doubtful, however, that either link in Dummett's chain will hold. I cannot now discuss the second link, the supposed convention of trying to say what is true. But it is relevant in the present context to comment on the claim that the utterance of an indicative sentence under conventional conditions constitutes an assertion.

One difficulty is obvious but may be superable: if there is to be a *general* account of assertion along these lines, there will have to be conventions that explain how assertions are made by uttering sentences not in the indicative mood. But perhaps it is plausible that if there are conventions linking indicatives and assertions, there are additional conventions linking other moods with assertions.

The trouble is that the required conventions do not exist. Of course it is true that if an indicative is uttered under the right conditions, an assertion will have been made. It may even be that we can specify conditions that are necessary and sufficient for making an assertion; for example, I think that in order to make an assertion a speaker must represent himself as believing what he says. But none of this suggests that the conditions are conventional in nature.

It must also be conceded that interpreters and speakers of a language are generally able to tell when an assertion has been made, and that this ability is an essential part of their linguistic competence. Furthermore, knowledge of linguistic and other conventions plays a key role in the making and detecting of assertions. Costume, stance, tone, office, role, and gesture have, or may have,

[7] Ibid., 302.

[8] For further discussion of Dummett's views on convention see Essay 18.

conventional aspects, and all these elements can make a crucial contribution to the force of an utterance. We may easily allow all this without agreeing that merely by following a convention, indicative or imperative utterances become assertions or commands.

There are, I think, strong reasons for rejecting the idea that making an assertion (or issuing a command, or asking a question) is performing a purely conventional action. One reason is, as I have been suggesting, that it is so hard to say what the convention is. (For example, if an asserter necessarily represents himself as believing what he says, one would have to describe the conventions by following which one can represent oneself as believing what one says.) A second point is this. Quite often we understand an utterance in all relevant respects except that we do not know whether it is an assertion. One kind of teasing consists in leaving the issue of assertion open in the mind of the teased; historical novels, or *romans-à-clef*, deliberately leave us puzzled. Is some conventional aspect of utterance omitted? What is it? And if we could say, then why would not the teaser or romancer include that very item in his utterance?

Whatever is conventional about assertion can be put into words, or somehow made an explicit part of the sentence. Let us suppose that this is not now the case, so that Frege's assertion sign is not just the formal equivalent of the indicative mood, but a more complete expression of the conventional element in assertion. It is easy to see that merely speaking a sentence in the strengthened mood cannot be counted on to result in an assertion; every joker, storyteller, and actor will immediately take advantage of the strengthened mood to simulate assertion. There is no point, then, in the strengthened mood; the available indicative does as well as language can do in the service of assertion. But since the indicative is not so strong that its mere employment constitutes assertion, what must be added to produce assertion cannot be merely a matter of linguistic convention.

What this argument illustrates is a basic trait of language, what may be called the autonomy of linguistic meaning. Once a feature of language has been given conventional expression, it can be used to serve many extra-linguistic ends; symbolic representation necessarily breaks any close tie with extra-linguistic purpose. Applied to the present case, this means that there cannot be a form of speech which, solely by dint of its conventional meaning, can be used only for a given purpose, such as making an assertion or asking a question.

The argument has a simple form: mood is not a conventional sign of assertion or command because nothing is, or could be, a conventional sign of assertion or command. The reason for this, it should be stressed, is not that the illocutionary force of a speech act is a purely mental, interior, or intentional aspect of the act.[9] Of course assertion or command must be intentional, as must meaning in the narrow sense. But it is part of the intention that the act should be interpreted as assertive or commanding, and therefore part of the intention that something publicly apparent should invite the appropriate interpretation.

It would be easy to become involved in a dispute about the extent to which a speaker's intention to perform an act that will be interpreted as being assertive must be realized before his act is correctly called an assertion. It is too much to insist that an assertion has been made only if it is actually interpreted as an assertion; it is too little to demand only that the intention be present. We need not settle the question how far an asserter must succeed in his intention; all that matters here is whether an asserter or commander must intend his hearer to recognize his intention through his (the asserter's) employment of what he knows or believes to be a linguistic convention. If there were such a convention, we should find it easy to say what it is, and easy, in the great majority of cases, to say whether or not it has been observed. But though we can usually determine whether or not an assertion has been made, we cannot in general say what convention was followed. The reason we cannot say is, I have urged, that there is no such convention.

It would be a mistake to conclude that there is no conventional connection between the moods and their uses. There would, indeed, be no such connection if certain analyses of the moods were correct. David Lewis, for example, has boldly suggested that all non-indicative sentences may be 'treated as paraphrases of the corresponding performatives, having the same base structure, meaning, intension and truth-value'.[10] Thus 'Fry that egg' would have the same analysis as 'I command that you fry that egg'. Lewis thinks the two sentences might have a different range of uses, but since this difference would not, on his theory, arise from a difference in

[9] I am indebted to Dummett for making me appreciate this point.
[10] D. Lewis, 'General Semantics', 208.

meaning, his theory simply denies that mood has any conventional significance.

An analysis with the same consequence was proposed many years ago by Herbert Bohnert.[11] Bohnert's proposal was that imperatives have the structure of disjunctions of a certain kind. Thus 'Fry that egg' would be rendered, 'Either you fry that egg, or X will happen', where X is something presumed unwanted by the person addressed.

These theories draw their strength from the fact that we can, and often do, use indicatives to do the work Dummett believes is conventionally assigned to the other moods. And in fact if we want to move in this direction, there is an even simpler and, I think, better, theory available, which is to assign to imperatives the same semantic analysis as is assigned to the most directly corresponding indicative (treat 'Fry that egg' just as 'You will fry that egg' is treated).[12]

It is a virtue of these theories that they make evident the fact that having a truth value is no obstacle to a sentence's being used to issue a command or ask a question. But this merit of reductive theories also accounts for their failure, for simply reducing imperatives or interrogatives to indicatives leaves us with no account at all of the differences among the moods. If any of the reductive theories is right, mood is as irrelevant to meaning as voice is often said to be. If mood does not affect meaning, how can we hope to explain the connection between mood and use, whatever the connection comes to? Reductive analyses abandon rather than solve the problem with which we began.

We are now in a position to list the characteristics a satisfactory theory of mood should have.

(1) It must show or preserve the relations between indicatives and corresponding sentences in the other moods; it must, for example, articulate the sense in which 'You will take off your shoes', 'Take off your shoes', and 'Will you take off your shoes?' have a common element.

[11] H. Bohnert, 'The Semiotic Status of Commands'.

[12] Yes–no interrogatives would then be treated, perhaps, as having the same semantics as the corresponding affirmative indicative; or, on another option, as having the same semantics as the alternation of the affirmative, with the negation of the affirmative, indicative. WH-questions might be assigned the same semantics as the corresponding open sentences in the indicative. Here as elsewhere in this article my remarks about the interrogative mood are sketchy. As Jaakko Hintikka has pointed out to me, my general programme for the moods may run into trouble when a serious attempt is made to apply it to interrogatives.

(2) It must assign an element of meaning to utterances in a given mood that is not present in utterances in other moods. And this element should connect with the difference in force between assertions, questions, and commands in such a way as to explain our intuition of a conventional relation between mood and use.

(3) Finally, the theory should be semantically tractable. If the theory conforms to the standards of a theory of truth, then I would say all is well. And on the other hand if, as I believe Bar-Hillel held, a standard theory of truth can be shown to be incapable of explaining mood, then truth theory is inadequate as a general theory of language.

The difficulty in meeting the three requirements is obvious. The first two conditions suggest that mood must be represented by operators that govern sentences, the sentences governed being either indicative (in which case no operator is needed for the indicative mood), or neutral (in which case an operator is needed for every mood). The third condition, however, seems to prohibit all but truth-functional sentential operators, and it is clear that truth-functional operators cannot serve to give a plausible interpretation of mood.

Dummett seems to me to be right when he says that the syntactic expression of mood is not like a functional expression, that we cannot ask what its sense or reference is, and that (therefore) a sentence with a mood indicator 'neither expresses anything nor stands for anything'. As Geach says, a mood indicator is not like any other part of speech; '. . . it is necessarily *sui generis*. For any other logical sign, if not superfluous, somehow modifies the content of a proposition; whereas this does not modify the content . . .'[13]

Dummett and Geach make these negative claims for what seem to me partly wrong, or confused, reasons. Dummett thinks a sentence with a mood indicator cannot express or stand for anything because the sentence '. . . asserts something, . . . namely, that the thought expressed by what follows the judgment-stroke is true'.[14] This is something I have urged that no expression can do; but the idea also seems wrong for another reason. If the assertion sign asserts that the thought expressed by the rest of the sentence is true, then the imperative sign should assert that the thought expressed by

[13] P. T. Geach, 'Assertion', 458.
[14] M. Dummett, *Frege: Philosophy of Language*, 316.

the rest of the sentence is to be made true. But this proposal wipes out the distinction between assertion and command. Geach says instead that the mood indicator (understood as Frege understood it) '. . . shows that the proposition is being asserted'.[15] This proposal preserves the needed distinction.

I have argued against both Geach and Dummett that no mood indicator can show or assert or in any other way conventionally determine what force its utterance has. But if this is so, we are left with no clear account of what mood contributes to meaning. Indeed, we seem to have a paradox. Mood must somehow contribute to meaning (point 2 above), since mood is clearly a conventional feature of sentences. Yet it cannot combine with or modify the meaning of the rest of the sentence in any known way.

Let us turn for help to what Austin called the 'explicit performatives'. We have rejected the idea put forward by David Lewis that imperatives be reduced to explicit performatives, but it remains open to exploit analogies. Austin drew attention to the fact that '. . . we can on occasion use the utterance "Go" to achieve practically the same as we achieve by the utterance "I order you to go"'.[16] But how are explicit performatives to be analysed?

Austin held that performatives have no truth value on the ground that uttering a sentence like 'I order you to go' is not typically to describe one's own speech act but rather to issue an order. This is perhaps an accurate account of how we would characterize many speech acts that consist in uttering explicit performatives. But as a description of what the words that are uttered mean, this view introduces an intolerable discrepancy between the semantics of certain first-person present-tense verbs and their other-person other-tense variants. And the problem is adventitious, since what is special about explicit performatives is better explained as due to a special use of words with an ordinary meaning than as due to a special meaning.

If we accept any of the usual semantics for explicit performatives, however, the difficulty recurs in a form that is hard to avoid. According to standard accounts of the matter, in a sentence like 'Jones ordered Smith to go' the final words ('Smith to go') serve to name or describe a sentence, or a proposition, or the sense of a

[15] P. T. Geach, 'Assertion', 458.

[16] J. L. Austin, *How to Do Things with Words*, 32.

sentence. To show the relevant embedded sentence, we may recast the whole thus: 'Jones ordered Smith to make it the case that Smith goes.' And now, on the standard accounts, the sentence 'Smith goes' cannot, in this context, have anything like its ordinary meaning. Therefore, neither can it have anything like its ordinary range of uses. However, 'I order you to go' (or, recast, 'I order you to make it the case that you go') has the same form as 'Jones ordered Smith to go', and so should have the same analysis with appropriate changes of person and time. It follows that in uttering 'I order you to go' I cannot mean by the words 'you go' anything like what I would mean by them if they stood alone; in the present context, I am using these words merely to refer to a sentence or the proposition it expresses. It seems quite impossible, then, that if any standard analysis of such sentences is correct, an utterance of 'I order you to go' could be an order to go.

Or, to make the same point with assertion: one way of establishing the fact that I am not asserting that it is raining when I utter the words 'It is raining' is by prefixing the words 'Jones asserted that'. According to most analyses of such sentences, the same effect should be expected if I prefix the words 'I assert that'.

This difficulty is one among the difficulties with the usual analyses that has prompted me to urge an entirely different approach to the semantics of indirect discourse, belief sentences, sentences about commands, orders, hopes, expectations, and so on: the whole unholy array of attitude-attributing locutions.[17] Leaving aside complications that arise when we quantify from outside into the governed sentences (or their scrambled surfaces), my proposal is this. Bearing in mind that it is in any case utterances, not sentences, that have a specific truth value and semantics, we should be satisfied with an analysis of the truth conditions of utterances of words like 'Jones asserted that it is raining'. I suggest that we view such an utterance as the utterance of two sentences; 'Jones asserted that', and then, 'It is raining'. If I assert that Jones asserted that it is raining, I do this by asserting 'Jones asserted that' and then uttering, usually non-assertively, the sentence that gives the content of Jones's assertion; in this case, 'It is raining'. The function of the 'that' in an utterance of 'Jones asserted that' is to refer to the following utterance, which gives the content. So to put the idea in a wordy but

[17] See Essay 7.

suggestive way: an utterance of 'Jones asserted that it is raining' has the effect of two utterances:

> Jones made an assertion whose content is given by my next utterance. It is raining.

This analysis accounts for the usual failure of substitutivity in attributions of attitude without invoking any non-standard semantics, for the reference of the 'that' changes with any change in the following utterance. It also allows the second utterance to consist, on occasion, in making an assertion, as it will if I say truly, 'I make an assertion whose content is given by my next utterance'. Similarly, I may be giving an order in saying 'You go' even if these words follow an utterance of 'I order that', or 'This is an order'.

I propose to treat the non-indicative moods in much the same way as explicit performatives, but not by reducing the other moods to the indicative. Here is the idea. Indicatives we may as well leave alone, since we have found no intelligible use for an assertion sign. We will go on, as is our wont, sometimes using indicative sentences to make assertions, sometimes using them to do other things; and we will continue to use sentences in other moods to make assertions when we can and find it fun.

In English we mark the non-indicative moods in various, occasionally ambiguous, ways, by changes in the verb, word order, punctuation, or intonation. We may think of non-indicative sentences, then, as indicative sentences plus an expression that syntactically represents the appropriate transformation; call this expression the *mood-setter*. And just as a non-indicative sentence may be decomposed into an indicative sentence and a mood-setter, so an utterance of a non-indicative sentence may be decomposed into two distinct speech acts, one the utterance of an indicative sentence, and the other the utterance of a mood-setter. It should not bother us that in fact we do not usually perform these acts one after the other but more or less simultaneously. Just think of someone rubbing his stomach with one hand and patting his head with the other.

We have seen that the mood-setter cannot be treated semantically as a sentential operator of any ordinary sort, and that it seems quite impossible to give a plausible account of how the meaning of a non-indicative sentence can be the result of combining the meaning of an indicative with the meaning of the mood-setter. I suggest that we accept the semantic independence of indicatives from their accompanying mood-setters by not trying to incorporate the mood-setter in a simple sentence in the indicative. There is the indicative sentence on the one hand, and before, or alongside, the mood-setter. Or better, thinking of the utterance, there is the utterance of the indicative element, and there is (perhaps simultaneously) the utterance of the mood-setter. The utterance of a non-indicative is thus always decomposable into the performance of two speech acts.

So far, the proposal is not clearly incompatible with the proposals of Geach, Dummett, and perhaps others. I have, indeed, dropped the assertion sign, but that may be considered largely a notational matter. I have also rejected an explanation of the meaning of the mood-operator in terms of a conventional indicator of the force with which the particular utterance is made. So there is a vacuum at the centre of my account; I have failed to say what the mood-setter means.

Geach remarked that what I call a mood-setter cannot be regarded as any other part of speech. This was because he thought of it as a part of a longer sentence, and realized it did not have the semantic properties of a sentential operator. We have removed the mood-setter from the indicative sentence it accompanies. The only form the mood-setter can have—the only function it can perform—is that of a sentence. It behaves like a sentence an utterance of which refers to an utterance of an indicative sentence. If we were to represent in linear form the utterance of, say, the imperative sentence 'Put on your hat', it would come out as the utterance of a sentence like 'My next utterance is imperatival in force', followed by an utterance of 'You will put on your hat'.

This suggests the semantic situation, but syntax makes it wrong. The mood-setter cannot be any actual sentence of English, since it represents a certain transformation. I do not want to claim that imperative sentences are two indicative sentences. Rather, we can give the semantics of the utterance of an imperative sentence by considering two specifications of truth conditions, the truth conditions of the utterance of an indicative sentence got by transforming the original imperative, and the truth conditions of the mood-setter. The mood-setter of an utterance of 'Put on your hat' is true if and only if the utterance of the indicative core is imperatival in force.

Mood-setters characterize an utterance as having a certain illocutionary force; they do not assert that it has that force, since only speakers make assertions. But if someone wishes to give an order, he may well do it by uttering the imperative mood-setter assertively. Then if the truth conditions of the mood-setter hold (if what the speaker has asserted is true), his utterance of the indicative core will constitute giving an order. There are plenty of other ways he can give the same order; for example, by asserting 'This is an order', or 'I hereby command that'; or simply by uttering 'You will take off your hat' as an order.

I believe this proposal satisfies the three requirements we listed for a satisfactory analysis of the moods.

First, on the proposal there is an element common to the moods. Syntactically, it is the indicative core, which is transformed in the non-indicative moods. Semantically, it is the truth conditions of this indicative core.

Second, mood is systematically represented by the mood-setter (or its absence in the case of the indicative). Mood-setters function semantically as sentences, utterances of which are true or false according as the utterance of the indicative core does or does not have the specified illocutionary force. The meaning of the mood-setter is conventional in whatever sense meaning in general is, but there is no suggestion that this meaning determines the illocutionary force of an utterance of the mood-setter, of its associated indicative, or of the pair. The conventional connection between mood and force is rather this: the concept of force is part of the meaning of mood. An utterance of an imperative sentence in effect says of itself that it has a certain force. But this is not the 'says' of 'asserts' (except on occasion and in addition). What it says, in this non-asserted sense, may as like be false as true. This fact does not affect the conceptual connection between mood and force.

Third, a straightforward semantics, based on a theory of truth for utterances, works as well here as elsewhere. In particular, all the utterances the theory takes as basic have a truth value in the standard sense. On the other hand, if I am right, the utterance of a non-indicative sentence cannot be said to have a truth value. For each utterance of a non-indicative has its mood-setter, and so must be viewed semantically as consisting in two utterances. Each of the two utterances has a truth value, but the combined utterance is not the utterance of a conjunction, and so does not have a truth value.

RADICAL INTERPRETATION

9 *Radical Interpretation*

Kurt utters the words 'Es regnet' and under the right conditions we know that he has said that it is raining. Having identified his utterance as intentional and linguistic, we are able to go on to interpret his words: we can say what his words, on that occasion, meant. What could we know that would enable us to do this? How could we come to know it? The first of these questions is not the same as the question what we *do* know that enables us to interpret the words of others. For there may easily be something we could know and don't, knowledge of which would suffice for interpretation, while on the other hand it is not altogether obvious that there is anything we actually know which plays an essential role in interpretation. The second question, how we could come to have knowledge that would serve to yield interpretations, does not, of course, concern the actual history of language acquisition. It is thus a doubly hypothetical question: given a theory that would make interpretation possible, what evidence plausibly available to a potential interpreter would support the theory to a reasonable degree? In what follows I shall try to sharpen these questions and suggest answers.

The problem of interpretation is domestic as well as foreign: it surfaces for speakers of the same language in the form of the question, how can it be determined that the language is the same? Speakers of the same language can go on the assumption that for them the same expressions are to be interpreted in the same way, but this does not indicate what justifies the assumption. All understanding of the speech of another involves radical interpretation. But it will help keep assumptions from going unnoticed to focus on cases where interpretation is most clearly called for: interpretation in one idiom of talk in another.[1]

What knowledge would serve for interpretation? A short answer would be, knowledge of what each meaningful expression means. In German, those words Kurt spoke mean that it is raining and Kurt was speaking German. So in uttering the words 'Es regnet', Kurt said that it was raining. This reply does not, as might first be thought, merely restate the problem. For it suggests that in passing from a description that does not interpret (his uttering of the words 'Es regnet') to interpreting description (his saying that it is raining) we must introduce a machinery of words and expressions (which may or may not be exemplified in actual utterances), and this suggestion is important. But the reply is no further help, for it does not say what it is to know what an expression means.

There is indeed also the hint that corresponding to each meaningful expression that is an entity, its meaning. This idea, even if not wrong, has proven to be very little help: at best it hypostasizes the problem.

Disenchantment with meanings as implementing a viable account of communication or interpretation helps explain why some philosophers have tried to get along without, not only meanings, but any serious theory at all. It is tempting, when the concepts we summon up to try to explain interpretation turn out to be more baffling than the explanandum, to reflect that after all verbal communication consists in nothing more than elaborate disturbances in the air which form a causal link between the non-linguistic activities of human agents. But although interpretable speeches are nothing but (that is, identical with) actions performed with assorted non-linguistic intentions (to warn, control, amuse, distract, insult), and these actions are in turn nothing but (identical with) intentional movements of the lips and larynx, this observation takes us no distance towards an intelligible general account of what we might know that would allow us to redescribe uninterpreted utterances as the right interpreted ones.

Appeal to meanings leaves us stranded further than we started from the non-linguistic goings-on that must supply the evidential

[1] The term 'radical interpretation' is meant to suggest strong kinship with Quine's 'radical translation'. Kinship is not identity, however, and 'interpretation' in place of 'translation' marks one of the differences: a greater emphasis on the explicitly semantical in the former.

base for interpretation; the 'nothing but' attitude provides no clue as to how the evidence is related to what it surely is evident for.

Other proposals for bridging the gap fall short in various ways. The 'causal' theories of Ogden and Richards and of Charles Morris attempted to analyse the meaning of sentences, taken one at a time, on the basis of behaviouristic data. Even if these theories had worked for the simplest sentences (which they clearly did not), they did not touch the problem of extending the method to sentences of greater complexity and abstractness. Theories of another kind start by trying to connect words rather than sentences with non-linguistic facts. This is promising because words are finite in number while sentences are not, and yet each sentence is no more than a concatenation of words: this offers the chance of a theory that interprets each of an infinity of sentences using only finite resources. But such theories fail to reach the evidence, for it seems clear that the semantic features of words cannot be explained directly on the basis of non-linguistic phenomena. The reason is simple. The phenomena to which we must turn are the extra-linguistic interests and activities that language serves, and these are served by words only in so far as the words are incorporated in (or on occasion happen to be) sentences. But then there is no chance of giving a foundational account of words before giving one of sentences.

For quite different reasons, radical interpretation cannot hope to take as evidence for the meaning of a sentence an account of the complex and delicately discriminated intentions with which the sentence is typically uttered. It is not easy to see how such an approach can deal with the structural, recursive feature of language that is essential to explaining how new sentences can be understood. But the central difficulty is that we cannot hope to attach a sense to the attribution of finely discriminated intentions independently of interpreting speech. The reason is not that we cannot ask necessary questions, but that interpreting an agent's intentions, his beliefs and his words are parts of a single project, no part of which can be assumed to be complete before the rest is. If this is right, we cannot make the full panoply of intentions and beliefs the evidential base for a theory of radical interpretation.

We are now in a position to say something more about what would serve to make interpretation possible. The interpreter must be able to understand any of the infinity of sentences the speaker might utter. If we are to state explicitly what the interpreter might know that would enable him to do this, we must put it in finite form.[2] If this requirement is to be met, any hope of a universal method of interpretation must be abandoned. The most that can be expected is to explain how an interpreter could interpret the utterances of speakers of a single language (or a finite number of languages): it makes no sense to ask for a theory that would yield an explicit interpretation for any utterance in any (possible) language.

It is still not clear, of course, what it is for a theory to yield an explicit interpretation of an utterance. The formulation of the problem seems to invite us to think of the theory as the specification of a function taking utterances as arguments and having interpretations as values. But then interpretations would be no better than meanings and just as surely entities of some mysterious kind. So it seems wise to describe what is wanted of the theory without apparent reference to meanings or interpretations: someone who knows the theory can interpret the utterances to which the theory applies.

The second general requirement on a theory of interpretation is that it can be supported or verified by evidence plausibly available to an interpreter. Since the theory is general—it must apply to a potential infinity of utterances—it would be natural to think of evidence in its behalf as instances of particular interpretations recognized as correct. And this case does, of course, arise for the interpreter dealing with a language he already knows. The speaker of a language normally cannot produce an explicit finite theory for his own language, but he can test a proposed theory since he can tell whether it yields correct interpretations when applied to particular utterances.

In radical interpretation, however, the theory is supposed to supply an understanding of particular utterances that is not given in advance, so the ultimate evidence for the theory cannot be correct sample interpretations. To deal with the general case, the evidence must be of a sort that would be available to someone who does not already know how to interpret utterances the theory is designed to cover: it must be evidence that can be stated without essential use of such linguistic concepts as meaning, interpretation, synonymy, and the like.

Before saying what kind of theory I think will do the trick, I want

[2] See Essay 1.

to discuss a last alternative suggestion, namely that a method of translation, from the language to be interpreted into the language of the interpreter, is all the theory that is needed. Such a theory would consist in the statement of an effective method for going from an arbitrary sentence of the alien tongue to a sentence of a familiar language; thus it would satisfy the demand for a finitely stated method applicable to any sentence. But I do not think a translation manual is the best form for a theory of interpretation to take.[3]

When interpretation is our aim, a method of translation deals with a wrong topic, a relation between two languages, where what is wanted is an interpretation of one (in another, of course, but that goes without saying since any theory is in some language). We cannot without confusion count the language used in stating the theory as part of the subject matter of the theory unless we explicitly make it so. In the general case, a theory of translation involves three languages: the object language, the subject language, and the metalanguage (the languages from and into which translation proceeds, and the language of the theory, which says what expressions of the subject language translate which expressions of the object language). And in this general case, we can know which sentences of the subject language translate which sentences of the object language without knowing what any of the sentences of either language mean (in any sense, anyway, that would let someone who understood the theory interpret sentences of the object language). If the subject language happens to be identical with the language of the theory, then someone who understands the theory can no doubt use the translation manual to interpret alien utterances; but this is because he brings to bear two things he knows and that the theory does not state: the fact that the subject language is his own, and his knowledge of how to interpet utterances in his own language.

It is awkward to try to make explicit the assumption that a mentioned sentence belongs to one's own language. We could try, for example, '"Es regnet" in Kurt's language is translated as "It is raining" in mine', but the indexical self-reference is out of place in a theory that ought to work for any interpreter. If we decide to accept

[3] The idea of a translation manual with appropriate empirical constraints as a device for studying problems in the philosophy of language is, of course, Quine's. This idea inspired much of my thinking on the present subject, and my proposal is in important respects very close to Quine's. Since Quine did not intend to answer the questions I have set, the claim that the method of translation is not adequate as a solution to the problem of radical interpretation is not a criticism of any doctrine of Quine's.

this difficulty, there remains the fact that the method of translation leaves tacit and beyond the reach of theory what we need to know that allows us to interpret our own language. A theory of translation must read some sort of structure into sentences, but there is no reason to expect that it will provide any insight into how the meanings of sentences depend on their structure.

A satisfactory theory for interpreting the utterances of a language, our own included, will reveal significant semantic structure: the interpretation of utterances of complex sentences will systematically depend on the interpretation of utterances of simpler sentences, for example. Suppose we were to add to a theory of translation a satisfactory theory of interpretation for our own language. Then we would have exactly what we want, but in an unnecessarily bulky form. The translation manual churns out, for each sentence of the language to be translated, a sentence of the translator's language; the theory of interpretation then gives the interpretation of these familiar sentences. Clearly the reference to the home language is superfluous; it is an unneeded intermediary between interpretation and alien idiom. The only expressions a theory of interpretation has to mention are those belonging to the language to be interpreted.

A theory of interpretation for an object language may then be viewed as the result of the merger of a structurally revealing theory of interpretation for a known language, and a system of translation from the unknown language into the known. The merger makes all reference to the known language otiose; when this reference is dropped, what is left is a structurally revealing theory of interpretation for the object language—couched, of course, in familiar words. We have such theories, I suggest, in theories of truth of the kind Tarski first showed how to give.[4]

What characterizes a theory of truth in Tarski's style is that it entails, for every sentence *s* of the object language, a sentence of the form:

s is true (in the object language) if and only if *p*.

Instances of the form (which we shall call T-sentences) are obtained by replacing '*s*' by a canonical description of *s*, and '*p*' by a translation of *s*. The important undefined semantical notion in the theory is that of *satisfaction* which relates sentences, open or closed,

[4] A. Tarski, 'The Concept of Truth in Formalized Languages'.

to infinite sequences of objects, which may be taken to belong to the range of the variables of the object language. The axioms, which are finite in number, are of two kinds: some give the conditions under which a sequence satisfies a complex sentence on the basis of the conditions of satisfaction of simpler sentences, others give the conditions under which the simplest (open) sentences are satisfied. Truth is defined for closed sentences in terms of the notion of satisfaction. A recursive theory like this can be turned into an explicit definition along familiar lines, as Tarski shows, provided the language of the theory contains enough set theory; but we shall not be concerned with this extra step.

Further complexities enter if proper names and functional expressions are irreducible features of the object language. A trickier matter concerns indexical devices. Tarski was interested in formalized languages containing no indexical or demonstrative aspects. He could therefore treat sentences as vehicles of truth; the extension of the theory to utterances is in this case trivial. But natural languages are indispensably replete with indexical features, like tense, and so their sentences may vary in truth according to time and speaker. The remedy is to characterize truth for a language relative to a time and a speaker. The extension to utterances is again straightforward.[5]

What follows is a defence of the claim that a theory of truth, modified to apply to a natural language, can be used as a theory of interpretation. The defence will consist in attempts to answer three questions:

1. It is reasonable to think that a theory of truth of the sort described can be given for a natural language?
2. Would it be possible to tell that such a theory was correct on the basis of evidence plausibly available to an interpreter with no prior knowledge of the language to be interpreted?
3. If the theory were known to be true, would it be possible to interpret utterances of speakers of the language?

The first question is addressed to the assumption that a theory of truth can be given for a natural language; the second and third questions ask whether such a theory would satisfy the further demands we have made on a theory of interpretation.

[5] For a discussion of how a theory of truth can handle demonstratives and how Convention T must be modified, see S. Weinstein, 'Truth and Demonstratives'.

1. *Can a theory of truth be given for a natural language?*

It will help us to appreciate the problem to consider briefly the case where a significant fragment of a language (plus one or two semantical predicates) is used to state its own theory of truth. According to Tarski's Convention T, it is a test of the adequacy of a theory that it entails all the T-sentences. This test apparently cannot be met without assigning something very much like a standard quantificational form to the sentences of the language, and appealing, in the theory, to a relational notion of satisfaction.[6] But the striking thing about T-sentences is that whatever machinery must operate to produce them, and whatever ontological wheels must turn, in the end a T-sentence states the truth conditions of a sentence using resources no richer than, because the same as, those of the sentence itself. Unless the original sentence mentions possible worlds, intensional entities, properties, or propositions, the statement of its truth conditions does not.

There is no equally simple way to make the analogous point about an alien language without appealing, as Tarski does, to an unanalysed notion of translation. But what we can do for our own language we ought to be able to do for another; the problem, it will turn out, will be to know that we are doing it.

The restriction imposed by demanding a theory that satisfies Convention T seems to be considerable: there is no generally accepted method now known for dealing, within the restriction, with a host of problems, for example, sentences that attribute attitudes, modalities, general causal statements, counterfactuals, attributive adjectives, quantifiers like 'most', and so on. On the other hand, there is what seems to me to be fairly impressive progress. To mention some examples, there is the work of Tyler Burge on proper names,[7] Gilbert Harman on 'ought',[8] John Wallace on mass terms and comparatives,[9] and there is my own work on attributions of attitudes and performatives,[10] on adverbs, events, and singular causal statements,[11] and on quotation.[12]

If we are inclined to be pessimistic about what remains to be done

[6] See J. Wallace, 'On the Frame of Reference', and Essay 3.
[7] T. Burge, 'Reference and Proper Names'.
[8] G. Harman, 'Moral Relativism Defended'.
[9] J. Wallace, 'Positive, Comparative, Superlative'.
[10] See Essays 7 and 8.
[11] See Essays 6–10 in *Essays on Actions and Events*.
[12] See Essay 6.

(or some of what has been done!), we should think of Frege's magnificent accomplishment in bringing what Dummett calls 'multiple generality' under control.[13] Frege did not have a theory of truth in Tarski's sense in mind, but it is obvious that he sought, and found, structures of a kind for which a theory of truth can be given.

The work of applying a theory of truth in detail to a natural language will in practice almost certainly divide into two stages. In the first stage, truth will be characterized, not for the whole language, but for a carefully gerrymandered part of the language. This part, though no doubt clumsy grammatically, will contain an infinity of sentences which exhaust the expressive power of the whole language. The second part will match each of the remaining sentences to one or (in the case of ambiguity) more than one of the sentences for which truth has been characterized. We may think of the sentences to which the first stage of the theory applies as giving the logical form, or deep structure, of all sentences.

2. *Can a theory of truth be verified by appeal to evidence available before interpretation has begun?*

Convention T says that a theory of truth is satisfactory if it generates a T-sentence for each sentence of the object language. It is enough to demonstrate that a theory of truth is empirically correct, then, to verify that the T-sentences are true (in practice, an adequate sample will confirm the theory to a reasonable degree). T-sentences mention only the closed sentences of the language, so the relevant evidence can consist entirely of facts about the behaviour and attitudes of speakers in relation to sentences (no doubt by way of utterances). A workable theory must, of course, treat sentences as concatenations of expressions of less than sentential length, it must introduce semantical notions like satisfaction and reference, and it must appeal to an ontology of sequences and the objects ordered by the sequences. All this apparatus is properly viewed as theoretical construction, beyond the reach of direct verification. It has done its work provided only it entails testable results in the form of T-sentences, and these make no mention of the machinery. A theory of truth thus reconciles the demand for a theory that articulates

[13] M. Dummett, *Frege: Philosophy of Language*.

grammatical structure with the demand for a theory that can be tested only by what it says about sentences.

In Tarski's work, T-sentences are taken to be true because the right branch of the biconditional is assumed to be a translation of the sentence truth conditions for which are being given. But we cannot assume in advance that correct translation can be recognized without pre-empting the point of radical interpretation; in empirical applications, we must abandon the assumption. What I propose is to reverse the direction of explanation: assuming translation, Tarski was able to define truth; the present idea is to take truth as basic and to extract an account of translation or interpretation. The advantages, from the point of view of radical interpretation, are obvious. Truth is a single property which attaches, or fails to attach, to utterances, while each utterance has its own interpretation; and truth is more apt to connect with fairly simple attitudes of speakers.

There is no difficulty in rephrasing Convention T without appeal to the concept of translation: an acceptable theory of truth must entail, for every sentence *s* of the object language, a sentence of the form: *s* is true if and only if *p*, where '*p*' is replaced by any sentence that is true if and only if *s* is. Given this formulation, the theory is tested by evidence that T-sentences are simply true; we have given up the idea that we must also tell whether what replaces '*p*' translates *s*. It might seem that there is no chance that if we demand so little of T-sentences, a theory of interpretation will emerge. And of course this would be so if we took the T-sentences in isolation. But the hope is that by putting appropriate formal and empirical restrictions on the theory as a whole, individual T-sentences will in fact serve to yield interpretations.[14]

We have still to say what evidence is available to an interpreter—evidence, we now see, that T-sentences are true. The evidence cannot consist in detailed descriptions of the speaker's beliefs and intentions, since attributions of attitudes, at least where subtlety is required, demand a theory that must rest on much the same evidence as interpretation. The interdependence of belief and meaning is evident in this way: a speaker holds a sentence to be true because of what the sentence (in his language) means, and because of what he believes. Knowing that he holds the sentence to be true, and knowing the meaning, we can infer his belief; given enough

[14] For essential qualifications, see footnote 11 of Essay 2.

information about his beliefs, we could perhaps infer the meaning. But radical interpretation should rest on evidence that does not assume knowledge of meanings or detailed knowledge of beliefs.

A good place to begin is with the attitude of holding a sentence true, of accepting it as true. This is, of course, a belief, but it is a single attitude applicable to all sentences, and so does not ask us to be able to make finely discriminated distinctions among beliefs. It is an attitude an interpreter may plausibly be taken to be able to identify before he can interpret, since he may know that a person intends to express a truth in uttering a sentence without having any idea *what* truth. Not that sincere assertion is the only reason to suppose that a person holds a sentence to be true. Lies, commands, stories, irony, if they are detected as attitudes, can reveal whether a speaker holds his sentences to be true. There is no reason to rule out other attitudes towards sentences, such as wishing true, wanting to make true, believing one is going to make true, and so on, but I am inclined to think that all evidence of this kind may be summed up in terms of holding sentences to be true.

Suppose, then, that the evidence available is just that speakers of the language to be interpreted hold various sentences to be true at certain times and under specified circumstances. How can this evidence be used to support a theory of truth? On the one hand, we have T-sentences, in the form:

(T) 'Es regnet' is true-in-German when spoken by x at time t if and only if it is raining near x at t.

On the other hand, we have the evidence, in the form:

(E) Kurt belongs to the German speech community and Kurt holds true 'Es regnet' on Saturday at noon and it is raining near Kurt on Saturday at noon.

We should, I think, consider (E) as evidence that (T) is true. Since (T) is a universally quantified conditional, the first step would be to gather more evidence to support the claim that:

(GE) $(x)(t)$ (if x belongs to the German speech community then (x holds true 'Es regnet' at t if and only if it is raining near x at t)).

The appeal to a speech community cuts a corner but begs no question: speakers belong to the same speech community if the same theories of interpretation work for them.

The obvious objection is that Kurt, or anyone else, may be wrong about whether it is raining near him. And this is of course a reason for not taking (E) as conclusive evidence for (GE) or for (T); and a reason not to expect generalizations like (GE) to be more than generally true. The method is rather one of getting a best fit. We want a theory that satisfies the formal constraints on a theory of truth, and that maximizes agreement, in the sense of making Kurt (and others) right, as far as we can tell, as often as possible. The concept of maximization cannot be taken literally here, since sentences are infinite in number, and anyway once the theory begins to take shape it makes sense to accept intelligible error and to make allowance for the relative likelihood of various kinds of mistake.[15]

The process of devising a theory of truth for an unknown native tongue might in crude outline go as follows. First we look for the best way to fit our logic, to the extent required to get a theory satisfying Convention T, on to the new language; this may mean reading the logical structure of first-order quantification theory (plus identity) into the language, not taking the logical constants one by one, but treating this much of logic as a grid to be fitted on to the language in one fell swoop. The evidence here is classes of sentences always held true or always held false by almost everyone almost all of the time (potential logical truths) and patterns of inference. The first step identifies predicates, singular terms, quantifiers, connectives, and identity; in theory, it settles matters of logical form. The second step concentrates on sentences with indexicals; those sentences sometimes held true and sometimes false according to discoverable changes in the world. This step in conjunction with the first limits the possibilities for interpreting individual predicates. The last step deals with the remaining sentences, those on which there is not uniform agreement, or whose held truth value does not depend systematically on changes in the environment.[16]

[15] For more on getting a 'best fit' see Essays 10–12.

[16] Readers who appreciate the extent to which this account parallels Quine's account of radical translation in Chapter 2 of *Word and Object* will also notice the differences: the semantic constraint in my method forces quantificational structure on the language to be interpreted, which probably does not leave room for indeterminacy of logical form; the notion of stimulus meaning plays no role in my method, but its place is taken by reference to the objective features of the world which alter in conjunction with changes in attitude towards the truth of sentences; the principle of charity, which Quine emphasizes only in connection with the identification of the (pure) sentential connectives, I apply across the board.

This method is intended to solve the problem of the interdependence of belief and meaning by holding belief constant as far as possible while solving for meaning. This is accomplished by assigning truth conditions to alien sentences that make native speakers right when plausibly possible, according, of course, to our own view of what is right. What justifies the procedure is the fact that disagreement and agreement alike are intelligible only against a background of massive agreement. Applied to language, this principle reads: the more sentences we conspire to accept or reject (whether or not through a medium of interpretation), the better we understand the rest, whether or not we agree about them.

The methodological advice to interpret in a way that optimizes agreement should not be conceived as resting on a charitable assumption about human intelligence that might turn out to be false. If we cannot find a way to interpret the utterances and other behaviour of a creature as revealing a set of beliefs largely consistent and true by our own standards, we have no reason to count that creature as rational, as having beliefs, or as saying anything.

Here I would like to insert a remark about the methodology of my proposal. In philosophy we are used to definitions, analyses, reductions. Typically these are intended to carry us from concepts better understood, or clear, or more basic epistemologically or ontologically, to others we want to understand. The method I have suggested fits none of these categories. I have proposed a looser relation between concepts to be illuminated and the relatively more basic. At the centre stands a formal theory, a theory of truth, which imposes a complex structure on sentences containing the primitive notions of truth and satisfaction. These notions are given application by the form of the theory and the nature of the evidence. The result is a partially interpreted theory. The advantage of the method lies not in its free-style appeal to the notion of evidential support but in the idea of a powerful theory interpreted at the most advantageous point. This allows us to reconcile the need for a semantically articulated structure with a theory testable only at the sentential level. The more subtle gain is that very thin evidence in support of each of a potential infinity of points can yield rich results, even with respect to the points. By knowing only the conditions under which speakers hold sentences true, we can come out, given a satisfactory theory, with an interpretation of each sentence. It remains to make good on this last claim. The theory itself at best gives truth conditions. What we need to show is that if such a theory satisfies the constraints we have specified, it may be used to yield interpretations.

3. *If we know that a theory of truth satisfies the formal and empirical criteria described, can we interpret utterances of the language for which it is a theory?*

A theory of truth entails a T-sentence for each sentence of the object language, and a T-sentence gives truth conditions. It is tempting, therefore, simply to say that a T-sentence 'gives the meaning' of a sentence. Not, of course, by naming or describing an entity that is a meaning, but simply by saying under what conditions an utterance of the sentence is true.

But on reflection it is clear that a T-sentence does not give the meaning of the sentence it concerns: the T-sentences does fix the truth value relative to certain conditions, but it does not say the object language sentence is true *because* the conditions hold. Yet if truth values were all that mattered, the T-sentence for 'Snow is white' could as well say that it is true if and only if grass is green or $2 + 2 = 4$ as say that it is true if and only if snow is white. We may be confident, perhaps, that no satisfactory theory of truth will produce such anomalous T-sentences, but this confidence does not license us to make more of T-sentences.

A move that might seem helpful is to claim that it is not the T-sentence alone, but the canonical proof of a T-sentence, that permits us to interpret the alien sentence. A canonical proof, given a theory of truth, is easy to construct, moving as it does through a string of biconditionals, and requiring for uniqueness only occasional decisions to govern left and right precedence. The proof does reflect the logical form the theory assigns to the sentence, and so might be thought to reveal something about meaning. But in fact we would know no more than before about how to interpret if all we knew was that a certain sequence of sentences was the proof, from some true theory, of a particular T-sentence.

A final suggestion along these lines is that we can interpret a particular sentence provided we know a correct theory of truth that deals with the language of the sentence. For then we know not only the T-sentence for the sentence to be interpreted, but we also 'know' the T-sentences for all other sentences; and of course, all the proofs.

Then we would see the place of the sentence in the language as a whole, we would know the role of each significant part of the sentence, and we would know about the logical connections between this sentence and others.

If we knew that a T-sentence satisfied Tarski's Convention T, we would know that it was true, and we could use it to interpret a sentence because we would know that the right branch of the biconditional translated the sentence to be interpreted. Our present trouble springs from the fact that in radical interpretation we cannot assume that a T-sentence satisfies the translation criterion. What we have been overlooking, however, is that we have supplied an alternative criterion: this criterion is that the totality of T-sentences should (in the sense described above) optimally fit evidence about sentences held true by native speakers. The present idea is that what Tarski assumed outright for each T-sentence can be indirectly elicited by a holistic constraint. If that constraint is adequate, each T-sentence will in fact yield an acceptable interpretation.

A T-sentence of an empirical theory of truth can be used to interpret a sentence, then, provided we also know the theory that entails it, and know that it is a theory that meets the formal and empirical criteria.[17] For if the constraints are adequate, the range of acceptable theories will be such that any of them yields some correct interpretation for each potential utterance. To see how it might work, accept for a moment the absurd hypothesis that the constraints narrow down the possible theories to one, and this one implies the T-sentence (T) discussed previously. Then we are justified in using this T-sentence to interpret Kurt's utterance of 'Es regnet' as his saying that it is raining. It is not likely, given the flexible nature of the constraints, that all acceptable theories will be identical. When all the evidence is in, there will remain, as Quine has emphasized, the trade-offs between the beliefs we attribute to a speaker and the interpretations we give his words. But the resulting indeterminacy cannot be so great but that any theory that passes the tests will serve to yield interpretations.

[17] See footnote 11 of Essay 2 and Essay 12.

10 *Belief and the Basis of Meaning*

Meaning and belief play interlocking and complementary roles in the interpretation of speech. By emphasizing the connection between our grounds for attributing beliefs to speakers, and our grounds for assigning meanings to their utterances, I hope to explain some problematic features both of belief and of meaning.

We interpret a bit of linguistic behaviour when we say what a speaker's words mean on an occasion of use. The task may be seen as one of redescription. We know that the words 'Es schneit' have been uttered on a particular occasion and we want to redescribe this uttering as an act of saying that it is snowing.[1] What do we need to know if we are to be in a position to redescribe speech in this way, that is, to interpret the utterances of a speaker? Since a competent interpreter can interpret any of a potential infinity of utterances (or so we may as well say), we cannot specify what he knows by listing cases. He knows, for example, that in uttering 'Es schneit' under certain conditions and with a certain intent, Karl has said that it is snowing; but there are endless further cases. What we must do then is state a finite theory from which particular interpretations follow. The theory may be used to describe an aspect of the interpreter's competence at understanding what is said. We may, if we please, also maintain that there is a mechanism in the interpreter that corresponds to the theory. If this means only that there is some mechanism or other that performs that task, it is hard to see how the claim can fail to be true.

Theory of interpretation is the business jointly of the linguist,

[1] I use the expression 'says that' in the present context in such a way that a speaker says (on a particular occasion) that it is snowing if and only if he utters words that (on that occasion) mean that it is snowing. So a speaker may say that it is snowing without *his* meaning, or asserting, that it is snowing.

psychologist and philosopher. Its subject matter is the behaviour of a speaker or speakers, and it tells what certain of their utterances mean. Finally, the theory can be used to describe what every interpreter knows, namely a specifiable infinite subset of the truths of the theory. In what follows, I shall say a little, and assume a lot, about the form a theory of interpretation can take. But I want to focus on the question how we can tell that any such theory is true.

One answer comes pat. The theory is true if its empirical implications are true; we can test the theory by sampling its implications for truth. In the present case, this means noticing whether or not typical interpretations a theory yields for the utterances of a speaker are correct. We agreed that any competent interpreter knows whether the relevant implications are true; so any competent interpreter can test a theory in this way. This does not mean, of course, that finding a true theory is trivial; it does mean that given a theory, testing it may require nothing arcane.

The original question, however, is how we know that a particular interpretation is correct, and our pat answer is not addressed to this question. An utterance can no doubt be interpreted by a correct theory, but if the problem is to determine when an interpretation is correct, it is no help to support the theory that yields it by giving samples of correct interpretations. There is an apparent impasse; we need the theory before we can recognize evidence on its behalf.

The problem is salient because uninterpreted utterances seem the appropriate evidential base for a theory of meaning. If an acceptable theory could be supported by such evidence, that would constitute conceptual progress, for the theory would be specifically semantical in nature, while the evidence would be described in non-semantical terms. An attempt to build on even more elementary evidence, say behaviouristic evidence, could only make the task of theory construction harder, though it might make it more satisfying. In any case, we can without embarrassment undertake the lesser enterprise.

A central source of trouble is the way beliefs and meanings conspire to account for utterances. A speaker who holds a sentence to be true on an occasion does so in part because of what he means, or would mean, by an utterance of that sentence, and in part because of what he believes. If all we have to go on is the fact of honest utterance, we cannot infer the belief without knowing the meaning, and have no chance of inferring the meaning without the belief.

Various strategies for breaking into this circle suggest themselves.

One is to find evidence for what words mean that is independent of belief. It would have to be independent of intentions, desires, regrets, wishes, approvals, and conventions too, for all of these have a belief component. Perhaps there are some who think it would be possible to establish the correctness of a theory of interpretation without knowing, or establishing, a great deal about beliefs, but it is not easy to imagine how it could be done.

Far more plausible is the idea of deriving a theory of interpretation from detailed information about the intentions, desires, and beliefs of speakers (or interpreters, or both). This I take to be the strategy of those who undertake to define or explain linguistic meaning on the basis of non-linguistic intentions, uses, purposes, functions, and the like: the traditions are those of Mead and Dewey, of Wittgenstein and Grice. This strategy will not meet the present need either, I think.

There can be nothing wrong, of course, with the methodological maxim that when baffling problems about meanings, reference, synonymy, and so on arise, we should remember that these concepts, like those of word, sentence, and language themselves, abstract away from the social transactions and setting which give them what content they have. Everyday linguistic and semantic concepts are part of an intuitive theory for organizing more primitive data, so only confusion can result from treating these concepts and their supposed objects as if they had a life of their own. But this observation cannot answer the question how we know when an interpretation of an utterance is correct. If our ordinary concepts suggest a confused theory, we should look for a better theory, not give up theorizing.

There can be no objection either to detailing the complicated and important relations between what a speaker's words mean and his non-linguistic intentions and beliefs. I have my doubts about the possibility of *defining* linguistic meaning in terms of non-linguistic intentions and beliefs, but those doubts, if not the sources of those doubts, are irrelevant to the present theme.

The present theme is the nature of the evidence for the adequacy of a theory of interpretation. The evidence must be describable in non-semantic, non-linguistic terms if it is to respond to the question we have set; it must also be evidence we can imagine the virgin investigator having without his already being in possession of the theory it is supposed to be evidence for. This is where I spy trouble. There is a principled, and not merely a practical, obstacle to verifying the existence of detailed, general and abstract beliefs and intentions, while being unable to tell what a speaker's words mean. We sense well enough the absurdity in trying to learn without asking him whether someone believes there is a largest prime, or whether he intends, by making certain noises, to get someone to stop smoking by that person's recognition that the noises were made with that intention. The absurdity lies not in the fact that it would be very hard to find out these things without language, but in the fact that we have no good idea how to set about authenticating the existence of such attitudes when communication is not possible.

This point is not happily stated by saying that our sophisticated beliefs and intentions and thoughts are like silent utterances. My claim is only that making detailed sense of a person's intentions and beliefs cannot be independent of making sense of his utterances. If this is so, then an inventory of a speaker's sophisticated beliefs and intentions cannot be the evidence for the truth of a theory for interpreting his speech behaviour.

Since we cannot hope to interpret linguistic activity without knowing what a speaker believes, and cannot found a theory of what he means on a prior discovery of his beliefs and intentions, I conclude that in interpreting utterances from scratch—in *radical* interpretation—we must somehow deliver simultaneously a theory of belief and a theory of meaning. How is this possible?

In order to make the problem sharp and simple enough for a relatively brief discussion, let me make a change in the description of the evidential base for a theory of interpretation. Instead of utterances of expressions, I want to consider a certain attitude towards expressions, an attitude that may or may not be evinced in actual utterances. The attitude is that of holding true, relativized to time. We may as well suppose we have available all that could be known of such attitudes, past, present, and future. Finally, I want to imagine that we can describe the external circumstances under which the attitudes hold or fail to hold. Typical of the sort of evidence available then would be the following: a speaker holds 'Es schneit' true when and only when it is snowing. I hope it will be granted that it is plausible to say we can tell when a speaker holds a sentence to be true without knowing what he means by the sentence, or what beliefs he holds about its unknown subject matter, or what detailed intentions do or might prompt him to utter it. It is often argued that we must assume that most of a speaker's utterances are of sentences he holds true: if this is right, the independent availability of the evidential base is assured. But weaker assumptions will do, since even the compulsive liar and the perennial kidder may be found out.

The problem, then, is this: we suppose we know what sentences a speaker holds true, and when, and we want to know what he means and believes. Perhaps we could crack the case if we knew enough about his beliefs and intentions, but there is no chance of this without prior access to a theory of interpretation. Given the interpretations, we could read off beliefs from the evidential base, but this assumes what we want to know.

I am struck by the analogy with a well-known problem in decision theory. Suppose an agent is indifferent between getting $5.00, and a gamble that offers him $11.00 if a coin comes up heads, and $0.00 if it comes up tails. We might explain (i.e., 'interpret') his indifference by supposing that money has a diminishing marginal utility for him: $5.00 is midway on his subjective value scale between $0.00 and $11.00. We arrive at this by assuming the gamble is worth the sum of the values of the possible outcomes as tempered by their likelihoods. In this case, we assume that heads and tails are equally likely. Unfortunately there is an equally plausible alternative explanation: since $5.00 obviously isn't midway in utility between $0.00 and $11.00, the agent must believe tails are more likely to come up than heads; if he thought heads and tails equally probable, he would certainly prefer the gamble, which would then be equal to a straight offer of $5.50.

The point is obvious. Choices between gambles are the result of two psychological factors, the relative values the chooser places on the outcomes, and the probability he assigns to those outcomes, conditional on his choice. Given the agent's beliefs (his subjective probabilities) it's easy to compute his relative values from his choices; given his values, we can infer his beliefs. But given only his choices, how can we work out both his beliefs and his values?

The problem is much like the problem of interpretation. The solution in the case of decision theory is neat and satisfying; nothing as good is available in the theory of meaning. Still, one can, I think, see the possibility of applying an analogous strategy. Simplified a bit, Frank Ramsey's proposal for coping with the problem of decision theory is this.[2] Suppose that there are two alternatives, getting $11.00 and getting $0.00, and that there is an event *E* such that the agent is

[2] F. P. Ramsey, 'Truth and Probability'.

indifferent between the following two gambles: Gamble One—if *E* happens the agent receives $11.00; if *E* fails to happen he gets $0.00. Gamble Two—if *E* happens he gets $0.00; if *E* fails to happen he gets $11.00. The agent's indifference between the gambles shows that he must judge that *E* is as likely to happen as not. For if he thought *E* more likely to occur than not, he would prefer the first gamble which promises him $11.00 if *E* occurs, and if he thought *E* more likely not to occur than to occur he would prefer the second gamble which pairs *E*'s non-occurrence with $11.00. This solves, for decision theory, the problem of how to separate out subjective probability from subjective utility, for once an event like *E* is discovered, it is possible to scale other values, and then to determine the subjective probabilities of all events.

In this version of decision theory, the evidential base is preferences between alternatives, some of them wagers; preference here corresponds to the attitude of holding true in the case of interpretation, as I put that problem. Actual choices in decision theory correspond to actual utterances in interpretation. The explanation of a particular preference involves the assignment of a comparative ranking of values and an evaluation of probabilities. Support for the explanation doesn't come from a new kind of insight into the attitudes and beliefs of the agent, but from more observations of preferences of the very sort to be explained. In brief, to explain (i.e., interpret) a particular choice or preference, we observe *other* choices or preferences; these will support a theory on the basis of which the original choice or preference can be explained. Attributions of subjective values and probabilities are part of the theoretical structure, and are convenient ways of summarizing facts about the structure of basic preferences; there is no way to test for them independently. Broadly stated, my theme is that we should think of meanings and beliefs as interrelated constructs of a single theory just as we already view subjective values and probabilities as interrelated constructs of decision theory.

One way of representing some of the explanatory facts about choice behaviour elicited by a theory of decision is to assign numbers to measure, say, the subjective values of outcomes to a particular agent. So we might assign the numbers 0, 1, and 2 as measures of the values to someone of receiving $0.00, $5.00, and $11.00 respectively. To the unwary this could suggest that for that agent $11.00 was worth twice as much as $5.00. Only by studying the underlying theory would the truth emerge that the assignment of numbers to measure utilities was unique up to a linear transformation, but not beyond. The numbers 2, 4, and 6 would have done as well in recording the facts, but 6 is not twice 4. The theory makes sense of comparisons of differences, but not of comparisons of absolute magnitudes. When we represent the facts of preference, utility, and subjective probability by assigning numbers, only some of the properties of numbers are used to capture the empirically justified pattern. Other properties of the numbers used may therefore be chosen arbitrarily, like the zero point and the unit in measuring utility or temperature.

The same facts may be represented by quite different assignments of numbers. In the interpretation of speech, introducing such supposed entities as propositions to be meanings of sentences or objects of belief may mislead us into thinking the evidence justifies, or should justify, a kind of uniqueness that it does not. In the case of decision theory, we can establish exactly which properties of numbers are relevant to the measurement of utility and which to the measurement of probability. Propositions being much vaguer than numbers, it is not clear to what extent they are overdesigned for their job.

There is not just an analogy between decision theory and interpretation theory, there is a connection. Seen from the side of decision theory, there is what Ward Edwards once dubbed the 'presentation problem' for empirical applications of decision theory. To learn the preferences of an agent, particularly among complex gambles, it is obviously necessary to describe the options in words. But how can the experimenter know what those words mean to the subject? The problem is not merely theoretical: it is well known that two descriptions of what the experimenter takes to be the same option may elicit quite different responses from a subject. We are up against a problem we discussed a moment ago in connection with interpretation: it is not reasonable to suppose we can interpret verbal behaviour without fine-grained information about beliefs and intentions, nor is it reasonable to imagine we can justify the attribution of preferences among complex options unless we can interpret speech behaviour. A radical theory of decision must include a theory of interpretation and cannot presuppose it.

Seen from the side of a theory of interpretation, there is the obvious difficulty in telling when a person accepts a sentence as true. Decision theory, and the common-sense ideas that stand behind it, help make a case for the view that beliefs are best understood in their role of rationalizing choices or preferences. Here we are considering only one special kind of belief, the belief that a sentence is true. Yet even in this case, it would be better if we could go behind the belief to a preference which might show itself in choice. I have no detailed proposal to make at the moment how this might, or should, be done. A first important step has been made by Richard Jeffrey.[3] He eliminates some troublesome confusions in Ramsey's theory by reducing the rather murky ontology of the theory, which dealt with events, options, and propositions to an ontology of propositions only. Preferences between propositions holding true then becomes the evidential base, so that the revised theory allows us to talk of degrees of belief in the truth of propositions, and the relative strength of desires that propositions be true. As Jeffrey points out, for the purposes of his theory, the objects of these various attitudes could as well be taken to be sentences. If this change is made, we can unify the subject matter of decision theory and theory of interpretation. Jeffrey assumes, of course, the sentences are understood by agent and theory builders in the same way. But the two theories may be united by giving up this assumption. The theory for which we should ultimately strive is one that takes as evidential base preferences between sentences—preferences that one sentence rather than another be true. The theory would then explain individual preferences of this sort by attributing beliefs and values to the agent, and meanings to his words.[4]

In this paper I shall not speculate further on the chances for an integrated theory of decision and interpretation; so I return to the problem of interpreting utterances on the basis of information about when, and under what external circumstances, the sentences they exemplify are held true. The central ideas in what I have said so far may be summarized: behavioural or dispositional facts that can be described in ways that do not assume interpretations, but on which a theory of interpretation can be based, will necessarily be a vector of meaning and belief. One result is that to interpret a particular utterance it is necessary to construct a comprehensive theory for the interpretation of a potential infinity of utterances. The evidence for the interpretation of a particular utterance will therefore have to be evidence for the interpretation of all utterances of a speaker or community. Finally, if entities like meanings, propositions, and objects of belief have a legitimate place in explaining speech

[3] R. Jeffrey, *The Logic of Decision*.

[4] For progress in developing such a theory, see my 'Toward a Unified Theory of Meaning and Action'.

behaviour, it is only because they can be shown to play a useful role in the construction of an adequate theory. There is no reason to believe in advance that these entities will be any help, and so it cannot be an independent goal of a theory or analysis to identify the meanings of expressions or the objects of belief.

The appreciation of these ideas, which we owe largely to Quine, represents one of the few real breakthroughs in the study of language. I have put things in my own way, but I think that the differences between us are more matters of emphasis than of substance. Much that Quine has written understandably concentrates on undermining misplaced confidence in the usefulness or intelligibility of concepts like those of analyticity, synonymy, and meaning. I have tried to accentuate the positive. Quine, like the rest of us, wants to provide a theory of interpretation. His animadversions on meanings are designed to discourage false starts; but the arguments in support of the strictures provide foundations for an acceptable theory.

I have accepted what I think is essentially Quine's picture of the problem of interpretation, and the strategy for its solution that I want to propose will obviously owe a great deal to him. There also will be some differences. One difference concerns the form the theory should take. Quine would have us produce a translation manual (a function, recursively given) that yields a sentence in the language of the interpreter for each sentence of the speaker (or more than one sentence in the case of ambiguity). To interpret a particular utterance one would give the translating sentence and specify the translation manual. In addition, it would be necessary to know exactly what information was preserved by a translation manual that met the empirical constraints: what was invariant, so to speak, from one acceptable translation manual to another.

I suggest making the theory explicitly semantical in character, in fact, that the theory should take the form of a theory of truth in Tarski's style.[5] In Tarski's style, but with modifications to meet present problems. For one thing, we are after a *theory* of truth where Tarski is interested in an explicit definition. This is a modification I will not discuss now: it mainly concerns the question how rich an ontology is available in the language in which the theory is given. Secondly, in order to accommodate the presence of demonstrative elements in natural language it is necessary to relativize the theory of truth to times

[5] A. Tarski, 'The Concept of Truth in Formalized Languages'.

and speakers (and possibly to some other things). The third modification is more serious and comes to the heart of the business under discussion. Tarski's Convention T demands of a theory of truth that it put conditions on some predicate, say 'is true', such that all sentences of a certain form are entailed by it. These are just those sentences with the familiar form: '"Snow is white" is true if and only if snow is white'. For the formalized languages that Tarski talks about, T-sentences (as we may call these theorems) are known by their syntax, and this remains true even if the object language and metalanguage are different languages and even if for quotation marks we substitute something more manageable. But in radical interpretation a syntactical test of the truth of T-sentences would be worthless, since such a test would presuppose the understanding of the object language one hopes to gain. The reason is simple: the syntactical test is merely meant to formalize the relation of synonymy or translation, and this relation is taken as unproblematic in Tarski's work on truth. Our outlook inverts Tarski's: we want to achieve an understanding of meaning or translation by assuming a prior grasp of the concept of truth. What we require, therefore, is a way of judging the acceptability of T-sentences that is not syntactical, and makes no use of the concepts of translation, meaning, or synonymy, but is such that acceptable T-sentences will in fact yield interpretations.

A theory of truth will be materially adequate, that is, will correctly determine the extension of the truth predicate, provided it entails, for each sentence *s* of the object language, a theorem of the form '*s* is true if and only if *p*' where '*s*' is replaced by a description of *s* and '*p*' is replaced by a sentence that is true if and only if *s* is. For purposes of interpretation, however, truth in a T-sentence is not enough. A theory of truth will yield interpretations only if its T-sentences state truth conditions in terms that may be treated as 'giving the meaning' of object language sentences. Our problem is to find constraints on a theory strong enough to guarantee that it can be used for interpretation.

There are constraints of a formal nature that flow from the demand that the theory be finitely axiomatized, and that it satisfy Convention T (as appropriately modified).[6] If the metalanguage is taken to contain ordinary quantification theory, it is difficult, if not impossible, to discover anything other than standard quantificational structures in

[6] See Essays 5 and 9.

the object language. This does not mean that anything whatever can be read into the object language simply by assuming it to be in the metalanguage; for example, the presence of modal operators in the metalanguage does not necessarily lead to a theory of truth for a modal object language.

A satisfactory theory cannot depart much, it seems, from standard quantificational structures or their usual semantics. We must expect the theory to rely on something very like Tarski's sort of recursive characterization of satisfaction, and to describe sentences of the object language in terms of familiar patterns created by quantification and cross-reference, predication, truth-functional connections, and so on. The relation between these semantically tractable patterns and the surface grammar of sentences may, of course, be very complicated.

The result of applying the formal constraints is, then, to fit the object language as a whole to the procrustean bed of quantification theory. Although this can no doubt be done in many ways if any, it is unlikely that the differences between acceptable theories will, in matters of logical form, be great. The identification of the semantic features of a sentence will then be essentially invariant: correct theories will agree on the whole about the quantificational structure to be assigned to a given sentence.

Questions of logical form being settled, the logical constants of quantification theory (including identity) will have been perforce discovered in the object language (well concealed, probably, beneath the surface). There remain the further primitive expressions to be interpreted. The main problem is to find a systematic way of matching predicates of the metalanguage to the primitive predicates of the object language so as to produce acceptable T-sentences. If the metalanguage predicates translate the object language predicates, things will obviously come out right; if they have the same extensions, this might be enough. But it would be foreign to our programme to use these concepts in stating the constraints: the constraints must deal only with sentences and truth. Still, it is easy to see how T-sentences for sentences with indexical features sharply limit the choice of interpreting predicates; for example the T-sentence for 'Das ist weiss' must have something like this form: 'For all speakers of German x and all times t "Das ist weiss" is true spoken by x at t if and only if the object demonstrated by x at t is white'. There may, as Quine has pointed out in his discussions of ontological relativity, remain room for alternative ontologies, and so for alternative systems for interpreting the predicates of the object language. I believe the range of acceptable theories of truth can be reduced to the point where all acceptable theories will yield T-sentences that we can treat as giving correct interpretations, by application of further reasonable and non-question-begging constraints. But the details must be reserved for another occasion.

Much more, obviously, must be said about the empirical constraints on the theory—the conditions under which a T-sentence may be accepted as correct. We have agreed that the evidential base for the theory will consist of facts about the circumstances under which speakers hold sentences of their language to be true. Such evidence, I have urged, is neutral as between meaning and belief and assumes neither. It now needs to be shown that such data can provide a test for the acceptability of T-sentences.

I propose that we take the fact that speakers of a language hold a sentence to be true (under observed circumstances) as prima-facie evidence that the sentence is true under those circumstances. For example, positive instances of 'Speakers (of German) hold "Es schneit" true when, and only when, it is snowing' should be taken to confirm not only the generalization, but also the T-sentence, '"Es schneit" is true (in German) for a speaker *x* at time *t* if and only if it is snowing at *t* (and near *x*)'.

Not all the evidence can be expected to point the same way. There will be differences from speaker to speaker, and from time to time for the same speaker, with respect to the circumstances under which a sentence is held true. The general policy, however, is to choose truth conditions that do as well as possible in making speakers hold sentences true when (according to the theory and the theory builder's view of the facts) those sentences are true. That is the general policy, to be modified in a host of obvious ways. Speakers can be allowed to differ more often and more radically with respect to some sentences than others, and there is no reason not to take into account the observed or inferred individual differences that may be thought to have caused anomalies (as seen by the theory).[7]

Building the theory cannot be a matter of deciding on an appropriate T-sentence for one sentence of the object language at a time; a pattern must be built up that preserves the formal constraints discussed above while suiting the evidence as well as may be. And of

[7] For more on such modifications, see Essay 11, and particularly D. Lewis, 'Radical Interpretation'.

course the fact that a theory does not make speakers universal holders of truths is not an inadequacy of the theory; the aim is not the absurd one of making disagreement and error disappear. The point is rather that widespread agreement is the only possible background against which disputes and mistakes can be interpreted. Making sense of the utterances and behaviour of others, even their most aberrant behaviour, requires us to find a great deal of reason and truth in them. To see too much unreason on the part of others is simply to undermine our ability to understand what it is they are so unreasonable about. If the vast amount of agreement on plain matters that is assumed in communication escapes notice, it's because the shared truths are too many and too dull to bear mentioning. What we want to talk about is what's new, surprising, or disputed.

A theory for interpreting the utterances of a single speaker, based on nothing but his attitudes towards sentences, would, we may be sure, have many equally eligible rivals, for differences in interpretation could be offset by appropriate differences in the beliefs attributed. Given a community of speakers with apparently the same linguistic repertoire, however, the theorist will strive for a single theory of interpretation: this will greatly narrow his practical choice of preliminary theories for each individual speaker. (In a prolonged dialogue, one starts perforce with a socially applicable theory, and refines it as evidence peculiar to the other speaker accumulates.)

What makes a social theory of interpretation possible is that we can construct a plurality of private belief structures: belief is built to take up the slack between sentences held true by individuals and sentences true (or false) by public standards. What is private about belief is not that it is accessible to only one person, but that it may be idiosyncratic. Attributions of belief are as publicly verifiable as interpretations, being based on the same evidence: if we can understand what a person says, we can know what he believes.

If interpretation is approached in the style I have been discussing, it is not likely that only one theory will be found satisfactory. The resulting indeterminacy of interpretation is the semantic counterpart of Quine's indeterminacy of translation. On my approach, the degree of indeterminacy will, I think, be less than Quine contemplates: this is partly because I advocate adoption of the principle of charity on an across-the-board basis, and partly because the uniqueness of quantificational structure is apparently assured if Convention T is satisfied. But in any case the question of indeterminacy is not central to the concerns of this paper. Indeterminacy of meaning or translation does not represent a failure to capture significant distinctions; it marks the fact that certain apparent distinctions are not significant. If there is indeterminacy, it is because when all the evidence is in, alternative ways of stating the facts remain open. An analogy from decision theory has already been noted: if the numbers 1, 2, and 3 capture the meaningful relations in subjective value between three alternatives, then the numbers -17, -2, and $+13$ do as well. Indeterminacy of this kind cannot be of genuine concern.

What is important is that if meaning and belief are interlocked as I have suggested, then the idea that each belief has a definite object, and the idea that each word and sentence has a definite meaning, cannot be invoked in describing the goal of a successful theory. For even if, contrary to what may reasonably be expected, there were no indeterminacy at all, entities such as meanings and objects of belief would be of no independent interest. We could, of course, invent such entities with a clear conscience if we were sure there were no permissible variant theories. But if we knew this, we would know how to state our theories without mention of the objects.

Theories of belief and meaning may require no exotic objects, but they do use concepts which set such theories apart from the physical and other non-psychological sciences: concepts like those of meaning and belief are, in a fundamental way, not reducible to physical, neurological, or even behaviouristic concepts. This irreducibility is not due, however, to the indeterminacy of meaning or translation, for if I am right, indeterminacy is important only for calling attention to how the interpretation of speech must go hand in hand with the interpretation of action generally, and so with the attribution of desires and beliefs. It is rather the methods we must invoke in constructing theories of belief and meaning that ensures the irreducibility of the concepts essential to those theories. Each interpretation and attribution of attitude is a move within a holistic theory, a theory necessarily governed by concern for consistency and general coherence with the truth, and it is this that sets these theories forever apart from those that describe mindless objects, or describe objects as mindless.[8]

[8] See Essay 11 of *Essays on Actions and Events*.

11 *Thought and Talk*

What is the connection between thought and language? The dependence of speaking on thinking is evident, for to speak is to express thoughts. This dependence is manifest in endless further ways. Someone who utters the sentence 'The candle is out' as a sentence of English must intend to utter words that are true if and only if an indicated candle is out at the time of utterance, and he must believe that by making the sounds he does he is uttering words that are true only under those circumstances. These intentions and beliefs are not apt to be dwelt on by the fluent speaker. But though they may not normally command attention, their absence would be enough to show he was not speaking English, and the absence of any analogous thoughts would show he was not speaking at all.

The issue lies on the other side: can there be thought without speech? A first and natural reaction is that there can be. There is the familiar, irksome experience of not being able to find the words to express one's ideas. On occasion one may decide that the editorial writer has put a point better than one could oneself. And there is Norman Malcolm's dog who, having chased a squirrel into the woods, barks up the wrong tree.[1] It is not hard to credit the dog with the belief that the squirrel is in that tree.

A definite, if feebler, intuition tilts the other way. It is possible to wonder whether the speaker who can't find the right words has a clear idea. Attributions of intentions and beliefs to dogs smack of anthropomorphism. A primitive behaviourism, baffled by the privacy of unspoken thoughts, may take comfort in the view that thinking is really 'talking to oneself'—silent speech.

[1] N. Malcolm, 'Thoughtless Brutes'.

Beneath the surface of these opposed tendencies run strong, if turgid, currents, which may help to explain why philosophers have, for the most part, preferred taking a stand on the issue to producing an argument. Whatever the reason, the question of the relationship between thought and speech seems seldom to have been asked for its own sake. The usual assumption is that one or the other, speech or thought, is by comparison easy to understand, and therefore the more obscure one (whichever that is) may be illuminated by analysing or explaining it in terms of the other.

The assumption is, I think, false: neither language nor thinking can be fully explained in terms of the other, and neither has conceptual priority. The two are, indeed, linked, in the sense that each requires the other in order to be understood; but the linkage is not so complete that either suffices, even when reasonably reinforced, to explicate the other. To make good this claim what is chiefly needed is to show how thought depends on speech, and this is the thesis I want to refine, and then to argue for.

We attribute a thought to a creature whenever we assertively employ a positive sentence the main verb of which is psychological—in English, 'believes', 'knows', 'hopes', 'desires', 'thinks', 'fears', are examples—followed by a sentence and preceded by the name or description of the creature. (A 'that' may optionally or necessarily follow the verb.) Some such sentences attribute states, others report events or processes: 'believes', 'thinks', and 'wants' report states, while 'came to believe', 'forgot', 'concluded', 'noticed', 'is proving' report events or processes. Sentences that can be used to attribute a thought exhibit what is often called, or analysed as, semantic intensionality, which means that the attribution may be changed from true to false, or false to true, by substitutions in the contained sentences that would not alter the truth value of the sentence in isolation.

I do not take for granted that if a creature has a thought, then we can, with resources of the kind just sketched, correctly attribute that thought to him. But thoughts so attributable at least constitute a good sample of the totality.

It is doubtful whether the various sorts of thought can be reduced to one, or even to a few: desire, knowledge, belief, fear, interest, to name some important cases, are probably logically independent to the extent that none can be defined using the others, even along with such further notions as truth and cause. Nevertheless, belief is central to all kinds of thought. If someone is glad that, or notices that, or remembers that, or knows that, the gun is loaded, then he must believe that the gun is loaded. Even to wonder whether the gun is loaded, or to speculate on the possibility that the gun is loaded, requires the belief, for example, that a gun is a weapon, that it is a more or less enduring physical object, and so on. There are good reasons for not insisting on any particular list of beliefs that are needed if a creature is to wonder whether a gun is loaded. Nevertheless, it is necessary that there be endless interlocked beliefs. The system of such beliefs identifies a thought by locating it in a logical and epistemic space.

Having a thought requires that there be a background of beliefs, but having a particular thought does not depend on the state of belief with respect to that very thought. If I consider going to a certain concert, I know I will be put to a degree of trouble and expense, and I have more complicated beliefs about the enjoyment I will experience. I will enjoy hearing Beethoven's Grosse Fuge, say, but only provided the performance achieves a reasonable standard, and I am able to remain attentive. I have the thought of going to the concert, but until I decide whether to go, I have no fixed belief that I will go; until that time, I merely entertain the thought.

We may say, summarizing the last two paragraphs, that a thought is defined by a system of beliefs, but is itself autonomous with respect to belief.

We usually think that having a language consists largely in being able to speak, but in what follows speaking will play only an indirect part. What is essential to my argument is the idea of an interpreter, someone who understands the utterances of another. The considerations to be put forward imply, I think, that a speaker must himself be an interpreter of others, but I shall not try to demonstrate that an interpreter must be a speaker, though there may be good reason to hold this. Perhaps it is worth pointing out that the notion of a language, or of two people speaking the same language does not seem to be needed here. Two speakers could interpret each other's utterances without there being, in any ordinary sense, a common language. (I do not want to deny that in other contexts the notion of a shared language may be very important.)

The chief thesis of this paper is that a creature cannot have thoughts unless it is an interpreter of the speech of another. This thesis does not imply the possibility of reduction, behaviouristic or otherwise, of thought to speech; indeed the thesis imputes no priority to language, epistemological or conceptual. The claim also falls short of similar claims in that it allows that there may be thoughts for which the speaker cannot find words, or for which there are no words.

Someone who can interpret an utterance of the English sentence 'The gun is loaded' must have many beliefs, and these beliefs must be much like the beliefs someone must have if he entertains the thought that the gun is loaded. The interpreter must, we may suppose, believe that a gun is a weapon, and that it is a more or less enduring physical object. There is probably no definite list of things that must be believed by someone who understands the sentence 'The gun is loaded,' but it is necessary that there be endless interlocked beliefs.

An interpreter knows the conditions under which utterances of sentences are true, and often knows that if certain sentences are true, others must be. For example, an interpreter of English knows that if 'The gun is loaded and the door is locked' is true, then 'The door is locked' is true. The sentences of a language have a location in the logical space created by the pattern of such relationships. Obviously the pattern of relations between sentences is very much like the pattern of relations between thoughts. This fact has encouraged the view that it is redundant to take both patterns as basic. If thoughts are primary, a language seems to serve no purpose but to express or convey thoughts; while if we take speech as primary, it is tempting to analyse thoughts as speech dispositions: as Sellars puts it, '. . . thinking at the distinctly human level . . . is essentially verbal activity'.[2] But clearly the parallel between the structure of thoughts and the structure of sentences provides no argument for the primacy of either, and only a presumption in favour of their interdependence.

We have been talking freely of thoughts, beliefs, meanings, and interpretations; or rather, freely using sentences that contain these words. But of course it is not clear what entities, or sorts of entities, there must be to make systematic sense of such sentences. However, talk apparently of thoughts and sayings does belong to a familiar mode of explanation of human behaviour and must be considered an organized department of common sense which may as well be called a theory. One way of examining the relation between thought and language is by inspecting the theory implicit in this sort of explanation.

Part of the theory deals with the teleological explanation of action. We wonder why a man raises his arm; an explanation might be that he wanted to attract the attention of a friend. This explanation would fail

[2] W. Sellars, 'Conceptual Change', 82.

if the arm-raiser didn't believe that by raising his arm he would attract the attention of his friend, so the complete explanation of his raising his arm, or at any rate a more complete explanation, is that he wanted to attract the attention of his friend *and* believed that by raising his arm he would attract his friend's attention. Explanation of this familiar kind has some features worth emphasizing. It explains what is relatively apparent—an arm-raising—by appeal to factors that are far more problematical: desires and beliefs. But if we were to ask for evidence that the explanation is correct, this evidence would in the end consist of more data concerning the sort of event being explained, namely further behaviour which is explained by the postulated beliefs and desires. Adverting to beliefs and desires to explain action is therefore a way of fitting an action into a pattern of behaviour made coherent by the theory. This does not mean, of course, that beliefs are nothing but patterns of behaviour, or that the relevant patterns can be defined without using the concepts of belief and desire. Nevertheless, there is a clear sense in which attributions of belief and desire, and hence teleological explanations of belief and desire, are supervenient on behaviour more broadly described.

A characteristic of teleological explanation not shared by explanation generally is the way in which it appeals to the concept of *reason*. The belief and desire that explain an action must be such that anyone who had that belief and desire would have a reason to act in that way. What's more, the descriptions we provide of desire and belief must, in teleological explanation, exhibit the rationality of the action in the light of the content of the belief and the object of the desire.

The cogency of a teleological explanation rests, as remarked, on its ability to discover a coherent pattern in the behaviour of an agent. Coherence here includes the idea of rationality both in the sense that the action to be explained must be reasonable in the light of the assigned desires and beliefs, but also in the sense that the assigned desires and beliefs must fit with one another. The methodological presumption of rationality does not make it impossible to attribute irrational thoughts and actions to an agent, but it does impose a burden on such attributions. We weaken the intelligibility of attributions of thoughts of any kind to the extent that we fail to uncover a consistent pattern of beliefs and, finally, of actions, for it is only against a background of such a pattern that we can identify thoughts. If we see a man pulling on both ends of a piece of string, we may decide he is fighting against himself, that he wants to move the string in incompatible directions. Such an explanation would require elaborate backing. No problem arises if the explanation is that he wants to break the string.

From the point of view of someone giving teleological explanations of the actions of another, it clearly makes no sense to assign priority either to desires or to beliefs. Both are essential to the explanation of behaviour, and neither is more directly open to observation than the other. This creates a problem, for it means that behaviour, which is the main evidential basis for attributions of belief and desire, is reckoned the result of two forces less open to public observation. Thus where one constellation of beliefs and desires will rationalize an action, it is always possible to find a quite different constellation that will do as well. Even a generous sample of actions threatens to leave open an unacceptably large number of alternative explanations.

Fortunately a more refined theory is available, one still firmly based on common sense: the theory of preference, or decision-making under uncertainty. The theory was first made precise by Frank Ramsey, though he viewed it as a matter of providing a foundation for the concept of probability rather than as a piece of philosophical psychology.[3] Ramsey's theory works by quantifying strength of preference and degree of belief in such a way as to make sense of the natural idea that in choosing a course of action we consider not only how desirable various outcomes are, but also how apt available courses of action are to produce those outcomes. The theory does not assume that we can judge degrees of belief or make numerical comparisons of value directly. Rather it postulates a reasonable pattern of preferences between courses of action, and shows how to construct a system of quantified beliefs and desires to explain the choices. Given the idealized conditions postulated by the theory, Ramsey's method makes it possible to identify the relevant beliefs and desires uniquely. Instead of talking of postulation, we might put the matter this way: to the extent that we can see the actions of an agent as falling into a consistent (rational) pattern of a certain sort, we can explain those actions in terms of a system of quantified beliefs and desires.

We shall come back to decision theory presently; now it is time to turn to the question of how speech is interpreted. The immediate aim of a theory of interpretation is to give the meaning of an arbitrary

[3] F. P. Ramsey, 'Truth and Probability'.

utterance by a member of a language community. Central to interpretation, I have argued, is a theory of truth that satisfies Tarski's Convention T (modified in certain ways to apply to a natural language). Such a theory may be taken as giving an interpretation of each sentence a speaker might utter. To belong to a speech community—to be an interpreter of the speech of others—one needs, in effect, to know such a theory, and to know that it is a theory of the right kind.[4]

A theory of interpretation, like a theory of action, allows us to redescribe certain events in a revealing way. Just as a theory of action can answer the question of what an agent is doing when he has raised his arm by redescribing the act as one of trying to catch his friend's attention, so a method of interpretation can lead to redescribing the utterance of certain sounds as an act of saying that snow is white. At this point, however, the analogy breaks down. For decision theory can also explain actions, while it is not at all clear how a theory of interpretation can explain a speaker's uttering the words 'Snow is white'. But this is, after all, to be expected, for uttering words is an action, and so must draw for its teleological explanation on beliefs and desires. Interpretation is not irrelevant to the teleological explanation of speech, since to explain why someone said something we need to know, among other things, his own interpretation of what he said, that is, what he believes his words mean in the circumstances under which he speaks. Naturally this will involve some of his beliefs about how others will interpret his words.

The interlocking of the theory of action with interpretation will emerge in another way if we ask how a method of interpretation is tested. In the end, the answer must be that it helps bring order into our understanding of behaviour. But at an intermediate stage, we can see that the attitude of *holding true* or *accepting as true*, as directed towards sentences, must play a central role in giving form to a theory. On the one hand, most uses of language tell us directly, or shed light on the question, whether a speaker holds a sentence to be true. If a speaker's purpose is to give information, or to make an honest assertion, then normally the speaker believes he is uttering a sentence true under the circumstances. If he utters a command, we may usually take this as showing that he holds a certain sentence (closely related to the sentence uttered) to be false; similarly for many cases of deceit.

[4] See Essays 9 and 10.

When a question is asked, it generally indicates that the questioner does not know whether a certain sentence is true; and so on. In order to infer from such evidence that a speaker holds a sentence true we need to know much about his desires and beliefs, but we do not have to know what his words mean.

On the other hand, knowledge of the circumstances under which someone holds sentences true is central to interpretation. We saw in the case of thoughts that although most thoughts are not beliefs, it is the pattern of belief that allows us to identify any thought; analogously, in the case of language, although most utterances are not concerned with truth, it is the pattern of sentences held true that gives sentences their meaning.

The attitude of holding a sentence to be true (under specified conditions) relates belief and interpretation in a fundamental way. We can know that a speaker holds a sentence to be true without knowing what he means by it or what belief it expresses for him. But if we know he holds the sentence true *and* we know how to interpret it, then we can make a correct attribution of belief. Symmetrically, if we know what belief a sentence held true expresses, we know how to interpret it. The methodological problem of interpretation is to see how, given the sentences a man accepts as true under given circumstances, to work out what his beliefs are and what his words mean. The situation is again similar to the situation in decision theory where, given a man's preferences between alternative courses of action, we can discern both his beliefs and his desires. Of course it should not be thought that a theory of interpretation will stand alone, for as we noticed, there is no chance of telling when a sentence is held true without being able to attribute desires and being able to describe actions as having complex intentions. This observation does not deprive the theory of interpretation of interest, but assigns it a place within a more comprehensive theory of action and thought.[5]

It is still unclear whether interpretation is required for a theory of action, which is the question we set ourselves to answer. What is certain is that all the standard ways of testing theories of decision or preference under uncertainty rely on the use of language. It is relatively simple to eliminate the necessity for verbal responses on the part of the

[5] The interlocking of decision theory and radical interpretation is explored also in Essay 10, in Essay 12 of *Essays on Thought and Action*, and in 'Toward a Unified Theory of Meaning and Action'.

subject: he can be taken to have expressed a preference by taking action, by moving directly to achieve his end, rather than by saying what he wants. But this cannot settle the question of what he has chosen. A man who takes an apple rather than a pear when offered both may be expressing a preference for what is on his left rather than his right, what is red rather than yellow, what is seen first, or judged more expensive. Repeated tests may make some readings of his actions more plausible than others, but the problem will remain how to determine when he judges two objects of choice to be identical. Tests that involve uncertain events—choices between gambles—are even harder to present without using words. The psychologist, sceptical of his ability to be certain how a subject is interpreting his instructions, must add a theory of verbal interpretation to the theory to be tested. If we think of all choices as revealing a preference that one sentence rather than another be true, the resulting total theory should provide an interpretation of sentences, and at the same time assign beliefs and desires, both of the latter conceived as relating the agent to sentences or utterances. This composite theory would explain all behaviour, verbal and otherwise.

All this strongly suggests that the attribution of desires and beliefs (and other thoughts) must go hand in hand with the interpretation of speech, that neither the theory of decision nor of interpretation can be successfully developed without the other. But it remains to say, in more convincing detail, why the attribution of thought depends on the interpretation of speech. The general, and not very informative, reason is that without speech we cannot make the fine distinctions between thoughts that are essential to the explanations we can sometimes confidently supply. Our manner of attributing attitudes ensures that all the expressive power of language can be used to make such distinctions. One can believe that Scott is not the author of *Waverley* while not doubting that Scott is Scott; one can want to be the discoverer of a creature with a heart without wanting to be the discoverer of a creature with a kidney. One can intend to bite into the apple in the hand without intending to bite into the only apple with a worm in it; and so forth. The intensionality we make so much of in the attribution of thoughts is very hard to make much of when speech is not present. The dog, we say, knows that its master is home. But does it know that Mr Smith (who is his master), or that the president of the bank (who is that same master), is home? We have no real idea how to settle, or make sense of, these questions. It is much harder to say, when speech is not present, how to distinguish universal thoughts from conjunctions of thoughts, or how to attribute conditional thoughts, or thoughts with, so to speak, mixed quantification ('He hopes that everyone is loved by someone').

These considerations will probably be less persuasive to dog lovers than to others, but in any case they do not constitute an argument. At best what we have shown, or claimed, is that unless there is behaviour that can be interpreted as speech, the evidence will not be adequate to justify the fine distinctions we are used to making in the attribution of thoughts. If we persist in attributing desires, beliefs or other attitudes under these conditions, our attributions and consequent explanations of actions will be seriously underdetermined in that many alternative systems of attribution, many alternative explanations, will be equally justified by the available data. Perhaps this is all we can say against the attribution of thoughts to dumb creatures; but I do not think so.

Before going on I want to consider a possible objection to the general line I have been pursuing. Suppose we grant, the objector says, that very complex behaviour not observed in infants and elephants is necessary if we are to find application for the full apparatus available for the attribution of thoughts. Still, it may be said, the sketch of how interpretation works does not show that this complexity must be viewed as connected with language. The reason is that the sketch makes too much depend on the special attitude of being thought true. The most direct evidence for the existence of this attitude is honest assertion. But then it would seem that we could treat as speech the behaviour of creatures that never did anything with language except make honest assertions. Some philosophers do dream of such dreary tribes; but would we be right to say they had a language? What has been lost to view is what may be called *the autonomy of meaning*. Once a sentence is understood, an utterance of it may be used to serve almost any extra-linguistic purpose. An instrument that could be put to only one use would lack autonomy of meaning; this amounts to saying it should not be counted as a language. So the complexity of behaviour needed to give full scope to attributions of thought need not, after all, have exactly the same complexity that allows, or requires, interpretation as a language.

I agree with the hypothetical objector that autonomy of meaning is essential to language; indeed it is largely this that explains why linguistic meaning cannot be defined or analysed on the basis of extra-linguistic intentions and beliefs. But the objector fails to distinguish between a language that *could* be used for only one purpose and one that *is* used for only one purpose. An instrument that could be used for only one purpose would not be language. But honest assertion alone might yield a theory of interpretation, and so a language that, though capable of more, might never be put to further uses. (As a practical matter, the event is unthinkable. Someone who knows under what conditions his sentences are socially true cannot fail to grasp, and avail himself of, the possibilities for dishonest assertion—or for joking, story-telling, goading, exaggerating, insulting, and all the rest of the jolly crew.)

A method of interpretation tells us that for speakers of English an utterance of 'It is raining' by a speaker x at time t is true if and only if it is raining (near x) at t. To be armed with this information, and to know that others know it, is to know what an utterance means independently of knowing the purposes that prompted it. The autonomy of meaning also helps to explain how it is possible, by the use of language, to attribute thoughts. Suppose someone utters assertively the sentence 'Snow is white'. Knowing the conditions under which such an utterance is true I can add, if I please, 'I believe that too', thus attributing a belief to myself. In this case we may both have asserted that snow is white, but sameness of force is not necessary to the self-attribution. The other may say with a sneer, expressing disbelief, 'Snow is white'—and I may again attribute a belief to myself by saying, 'But *I* believe that'. It can work as well in another way: If I can take advantage of an utterance of someone else's to attribute a belief to myself, I can use an utterance of my own to attribute a belief to someone else. First I utter a sentence, perhaps 'Snow is white', and then I add 'He believes that'. The first utterance may or may not be an assertion; in any case, it does not attribute a belief to anyone (though if it is an assertion, then I do *represent* myself as believing that snow is white). But if my remark 'He believes that' is an assertion, I have attributed a belief to someone else. Finally, there is no bar to my attributing a belief to myself by saying first, 'Snow is white' and then adding, 'I believe that'.

In all these examples, I take the word 'that' to refer demonstratively to an utterance, whether it is an utterance by the speaker of the 'that' or by another speaker. The 'that' cannot refer to a sentence, both because, as Church has pointed out in similar cases, the reference would then have to be relativized to a language, since a sentence may have different meanings in different languages;[6] but also, and more obviously, because the same sentence may have different truth values in the same language.

What demonstrative reference to utterances does in the sort of case just considered it can do as well when the surface structure is altered to something like 'I believe that snow is white' or 'He believes that snow is white'. In these instances also I think we should view the 'that' as a demonstrative, now referring ahead to an utterance on the verge of production. Thus the logical form of standard attributions of attitude is that of two utterances paratactically joined. There is no connective, though the first utterance contains a reference to the second. (Similar remarks go, of course, for inscriptions of sentences.)

I have discussed this analysis of verbal attributions of attitude elsewhere, and there is no need to repeat the arguments and explanations here.[7] It is an analysis with its own difficulties, especially when it comes to analysing quantification into the contained sentence, but I think these difficulties can be overcome while preserving the appealing features of the idea. Here I want to stress a point that connects the paratactic analysis of attributions of attitude with our present theme. The proposed analysis directly relates the autonomous feature of meaning with our ability to describe and attribute thoughts, since it is only because the interpretation of a sentence is independent of its use that the utterance of a sentence can serve in the description of the attitudes of others. If my analysis is right, we can dispense with the unlikely (but common) view that a sentence bracketed into a 'that'-clause needs an entirely different interpretation from the one that works for it in other contexts. Since sentences are not names or descriptions in ordinary contexts, we can in particular reject the assumption that the attitudes have objects such as propositions which 'that'-clauses might be held to name or describe. There should be no temptation to call the utterance to which reference is made according to the paratactic analysis the object of the attributed attitude.

Here a facile solution to our problem about the relation between thoughts and speech suggests itself. One way to view the paratactic analysis, a way proposed by Quine in *Word and Object*, is this: when a speaker attributes an attitude to a person, what he does is ape or mimic an actual or possible speech act of that person.[8] Indirect discourse is

[6] A. Church, 'On Carnap's Analysis of Statements of Assertion and Belief'.
[7] See Essay 7.
[8] W. V. Quine, *Word and Object*, 219.

the best example, and assertion is another good one. Suppose I say, 'Herodotus asserted that the Nile rises in the Mountains of the Moon.' My second utterance—my just past utterance of 'The Nile rises in the Mountains of the Moon'—must, if my attribution to Herodotus is correct, bear a certain relationship to an utterance of Herodotus': it must, in some appropriate sense, be a translation of it. Since, assuming still that the attribution is correct, Herodotus and I are *samesayers*, my utterance mimicked his. Not with respect to force, of course, since I didn't assert anything about the Nile. The sameness is with respect to the content of our utterances. If we turn to other attitudes, the situation is more complicated, for there is typically no utterance to ape. If I affirm 'Jones believes that snow is white', my utterance of 'Snow is white' may have no actual utterance of Jones's to imitate. Still, we could take the line that what I affirm is that Jones would be honestly speaking his mind were he to utter a sentence translating mine. Given some delicate assumptions about the conditions under which such a subjunctive conditional is true, we could conclude that only someone with a language could have a thought, since to have a thought would be to have a disposition to utter certain sentences with appropriate force under given circumstances.

We could take this line, but unfortunately there seems no clear reason why we have to. We set out to find an argument to show that only creatures with speech have thoughts. What has just been outlined is not an argument, but a proposal, and a proposal we need not accept. The paratactic analysis of the logical form of attributions of attitude can get along without the mimic-theory of utterance. When I say, 'Jones believes that snow is white' I describe Jones's state of mind directly: it is indeed the state of mind someone is in who could honestly assert 'Snow is white' if he spoke English, but that may be a state a languageless creature could also be in.

In order to make my final main point, I must return to an aspect of interpretation so far neglected. I remarked that the attitude of holding true, directed to sentences under specified circumstances, is the basis for interpretation, but I did not say how it can serve this function. The difficulty, it will be remembered, is that a sentence is held true because of two factors: what the holder takes the sentence to mean, and what he believes. In order to sort things out, what is needed is a method for holding one factor steady while the other is studied.

Membership in a language community depends on the ability to interpret the utterances of members of the group, and a method is at hand if one has, and knows one has, a theory which provides truth conditions, more or less in Tarski's style, for all sentences (relativized, as always, to time and speaker). The theory is correct as long as it entails, by finitely stated means, theorems of the familiar form: ' "It is raining" is true for a speaker *x* at time *t* if and only if it is raining (near *x*) at *t*'. The evidential basis for such a theory concerns sentences held true, facts like the following: ' "It is raining" was held true by Smith at 8 a.m. on 26 August and it did rain near Smith at that time.' It would be possible to generate a correct theory simply by considering sentences to be true when held true, provided (1) there was a theory which satisfied the formal constraints and was consistent in this way with the evidence, and (2) all speakers held a sentence to be true just when that sentence was true—provided, that is, all beliefs, at least as far as they could be expressed, were correct.

But of course it cannot be assumed that speakers never have false beliefs. Error is what gives belief its point. We can, however, take it as given that *most* beliefs are correct. The reason for this is that a belief is identified by its location in a pattern of beliefs; it is this pattern that determines the subject matter of the belief, what the belief is about. Before some object in, or aspect of, the world can become part of the subject matter of a belief (true or false) there must be endless true beliefs about the subject matter. False beliefs tend to undermine the identification of the subject matter; to undermine, therefore, the validity of a description of the belief as being about that subject. And so, in turn, false beliefs undermine the claim that a connected belief is false. To take an example, how clear are we that the ancients—some ancients—believed that the earth was flat? *This* earth? Well, this earth of ours is part of the solar system, a system partly identified by the fact that it is a gaggle of large, cool, solid bodies circling around a very large, hot star. If someone believes *none* of this about the earth, is it certain that it is the earth that he is thinking about? An answer is not called for. The point is made if this kind of consideration of related beliefs can shake one's confidence that the ancients believed the earth was flat. It isn't that any one false belief necessarily destroys our ability to identify further beliefs, but that the intelligibility of such identifications must depend on a background of largely unmentioned and unquestioned true beliefs. To put it another way: the more things a believer is right about, the sharper his errors are. Too much mistake simply blurs the focus.

What makes interpretation possible, then, is the fact that we can dismiss a priori the chance of massive error. A theory of interpretation cannot be correct that makes a man assent to very many false sentences: it must generally be the case that a sentence is true when a speaker holds it to be. So far as it goes, it is in favour of a method of interpretation that it counts a sentence true just when speakers hold it to be true. But of course, the speaker may be wrong; and so may the interpreter. So in the end what must be counted in favour of a method of interpretation is that it puts the interpreter in general agreement with the speaker: according to the method, the speaker holds a sentence true under specified conditions, and these conditions obtain, in the opinion of the interpreter, just when the speaker holds the sentence to be true.

No simple theory can put a speaker and interpreter in perfect agreement, and so a workable theory must from time to time assume error on the part of one or the other. The basic methodological precept is, therefore, that a good theory of interpretation maximizes agreement. Or, given that sentences are infinite in number, and given further considerations to come, a better word might be *optimize*.

Some disagreements are more destructive of understanding than others, and a sophisticated theory must naturally take this into account. Disagreement about theoretical matters may (in some cases) be more tolerable than disagreement about what is more evident; disagreement about how things look or appear is less tolerable than disagreement about how they are; disagreement about the truth of attributions of certain attitudes to a speaker by that same speaker may not be tolerable at all, or barely. It is impossible to simplify the considerations that are relevant, for everything we know or believe about the way evidence supports belief can be put to work in deciding where the theory can best allow error, and what errors are least destructive of understanding. The methodology of interpretation is, in this respect, nothing but epistemology seen in the mirror of meaning.

The interpreter who assumes his method can be made to work for a language community will strive for a theory that optimizes agreement throughout the community. Since easy communication has survival value, he may expect usage within a community to favour simple common theories of interpretation.

If this account of radical interpretation is right, at least in broad outline, then we should acknowledge that the concepts of objective truth, and of error, necessarily emerge in the context of interpretation. The distinction between a sentence being held true and being in fact true is essential to the existence of an interpersonal system of communication, and when in individual cases there is a difference, it must be counted as error. Since the attitude of holding true is the same, whether the sentence is true or not, it corresponds directly to belief. The concept of belief thus stands ready to take up the slack between objective truth and the held true, and we come to understand it just in this connection.

We have the idea of belief only from the role of belief in the interpretation of language, for as a private attitude it is not intelligible except as an adjustment to the public norm provided by language. It follows that a creature must be a member of a speech community if it is to have the concept of belief. And given the dependence of other attitudes on belief, we can say more generally that only a creature that can interpret speech can have the concept of a thought.

Can a creature have a belief if it does not have the concept of belief? It seems to me it cannot, and for this reason. Someone cannot have a belief unless he understands the possibility of being mistaken, and this requires grasping the contrast between truth and error—true belief and false belief. But this contrast, I have argued, can emerge only in the context of interpretation, which alone forces us to the idea of an objective, public truth.

It is often wrongly thought that the semantical concept of truth is redundant, that there is no difference between asserting that a sentence *s* is true, and using *s* to make an assertion. What may be right is a redundancy theory of belief, that to believe that *p* is not to be distinguished from the belief that *p* is true. This notion of truth is not the semantical notion: language is not directly in the picture. But it is only just out of the picture; it is part of the frame. For the notion of a true belief depends on the notion of a true utterance, and this in turn there cannot be without shared language. As Shakespeare's Ulysses puts it:

> . . . no man is the lord of anything,
> Though in and of him there be much consisting,
> Till he communicate his parts to others;
> Nor doth he of himself know them for aught
> Till he behold them formed in th'applause
> Where they're extended.

(*Troilus and Cressida*, III. iii. 115–20.)

12 *Reply to Foster*

There is much with which I agree, and more I admire, in Mr Foster's paper. I share his bias in favour of extensional first-order languages; I am glad to keep him company in the search for an explicitly semantical theory that recursively accounts for the meanings of sentences in terms of their structures; and I am happy he concurs in holding that a theory may be judged adequate on the basis of holistic constraints. I especially applaud Foster for what he passes over: just as *Lear* gains power through the absence of Cordelia, I think treatments of language prosper when they avoid uncritical evocation of the concepts of convention, linguistic rule, linguistic practice, or language games.

Still on the positive side, I think Foster is right in asking whether a proposed theory explicitly states something knowledge of which would suffice for interpreting utterances of speakers of the language to which it applies. (I avoid the word 'mastery', and the special competence of a speaker, if any, for reasons that will not, I believe, affect our discussion.) I was slow to appreciate the importance of this way of formulating a general aim of theories of meaning, though elements of the idea appear in several early papers of mine.[1] I am grateful to a number of Oxford friends for prompting me to try to clarify my views on this subject—and here I should especially mention Michael Dummett, Gareth Evans, John McDowell, and John Foster.

In a paper first read in Biel, Switzerland, in May 1973, I criticized my own earlier attempts to say exactly what the relation is between a theory of truth and a theory of meaning, and I tried to do better.[2] I read this paper again in Windsor (November 1973), and it became the basis

[1] For example, Essays 1 and 2.
[2] The paper mentioned is Essay 9.

for much discussion in a seminar Michael Dummett and I gave in Oxford in Trinity Term, 1974. The criticisms I there levelled against my earlier formulation are (I believe) essentially those elaborated by Foster in the second part of his present paper, and my attempt at something better is among the views he attacks in the third part of his paper.

I am in general agreement with Foster that I have yet to give a completely satisfactory formulation of what it is, on my approach, that it suffices to know in order to be able to interpret a speaker's utterances. On the other hand, I hope I am not as far off target as he thinks, and I am not persuaded by his arguments that my 'grand design is in ruins'. Indeed it still seems to me right, as far as it goes, to hold that someone is in a position to interpret the utterances of speakers of a language *L* if he has a certain body of knowledge entailed by a theory of truth for *L*—a theory that meets specified empirical and formal constraints—and he knows that this knowledge is entailed by such a theory.

Tarski says, nearly enough for our purposes, that a theory of truth for a language *L* is satisfactory provided it entails (by a finite set of non-logical axioms and normal logic), for each sentence *s* of *L*, a theorem of the form:

s is true-in-*L* if and only if *p*

where '*s*' is replaced by a standardized description of *s* and '*p*' is replaced by a translation of *s* into the language of the theory. If we knew such a theory, and that it was such a theory, then we could produce a translation of each sentence of *L*, and would know that it was a translation. We would know more, for we would know in detail how the truth values of sentences of *L* were owed to their structures, and why some sentences entailed others, and how words performed their functions by dint of relations to objects in the world.

Since Tarski was interested in defining truth, and was working with artificial languages where stipulation can replace illumination, he could take the concept of translation for granted. But in *radical* interpretation, this is just what cannot be assumed. So I have proposed instead some empirical constraints on accepting a theory of truth that can be stated without appeal to such concepts as those of meaning, translation, or synonomy, though not without a certain understanding of the notion of truth. By a course of reasoning, I have tried to show that if the constraints are met by a theory, then the T-sentences that flow from that theory will in fact have translations of *s* replacing '*p*'.

To accept this change in perspective is not to give up Convention T but to read it in a new way. Like Tarski, I want a theory that satisfies Convention T, but where he assumes the notion of translation in order to throw light on that of truth, I want to illuminate the concept of translation by assuming a partial understanding of the concept of truth.

That empirical restrictions must be added to the formal restrictions if acceptable theories of truth are to include only those that would serve for interpretation was clear to me even when I wrote 'Truth and Meaning'. My mistake was not, as Foster seems to suggest, to suppose that *any* theory that correctly gave truth conditions would serve for interpretation; my mistake was to overlook the fact that someone might know a sufficiently unique theory without knowing that it was sufficiently unique. The distinction was easy for me to neglect because I imagined the theory to be known by someone who had constructed it from the evidence, and such a person could not fail to realize that his theory satisfied the constraints.

Foster notes the difference between two questions that might be raised about my proposal. One is, whether the constraints I have placed on an acceptable theory are adequate to ensure that it satisfies Convention T—i.e. to ensure that in its T-sentence, the right branch of the biconditional really does translate the sentence whose truth value it is giving. The other question is whether I have succeeded in saying what a competent interpreter knows (or what it would suffice for him to know). Foster is concerned here only with the second question; he is willing to grant, for the space of the argument, that the constraints are adequate to their purpose.

It is in this light that we must understand Foster's discussion of theories of truth that correctly give the extension of the truth predicate—theories all of whose T-sentences are true—but which do not satisfy Convention T. Thus Foster is going along with me (for the moment) in supposing that my criteria will not allow a theory that contains as a T-sentence the following:

> '*a* is part of *b*' is true if and only if *a* is a part of *b* and the Earth moves.

Foster's point is rather that although my interpreter has a theory that satisfies Convention T, nothing in the theory itself tells him this.

The same point comes up when Foster says,

> as ordinarily understood, to state the truth conditions of a sentence is to say what . . . is necessary and sufficient for its truth, to demarcate, within the total range of possible circumstances, that subset with which the sentence accords. But this is not the sense in which a *T-sentence* states truth conditions. A T-sentence does not say that such and such a structural type *would be* true . . . in all and only circumstances in which it *was* the case that . . ., but merely that, things being as they are, this structural type *is* true if and only if . . .

A theory that passes the empirical tests is one that in fact can be projected to unobserved and counterfactual cases, and this is apparent to anyone who knows what the evidence is and how it is used to support the theory. The trouble is, the theory does not state that it has the character it does.

We get a precise parallel if we ask what someone must know to be a physicist. A quick answer might be: the laws of physics. But Foster would say, and I agree, that this is not enough. The physicist must also know (and here I speak for myself) that those laws *are* laws—i.e. that they are confirmed by their instances, and support counterfactual and subjunctive claims. To get the picture, you are to imagine that a budding scientist is told that the mass of a body has no influence on how long it will take for it to fall a given distance in a vacuum. Then he is asked, 'Suppose Galileo had dropped a feather and a cannon ball from the top of the Empire State Building, and that the earth had no atmosphere. Which would have reached the ground sooner, the feather or the cannon ball?' The wise child replies, 'I have no idea. You told me only what *does* happen, things being as they are; you did not say what would happen if things were otherwise.'

Foster offers, as a thesis he thinks I may have 'tried to convey' but failed to get right, the following: 'what we need to know, for the mastery of *L*, are both the facts which [a T-theory] states, and that those facts as known by us, are T-theoretical.' He then puts it in a nutshell: what someone needs to know is that some T-theory for *L* states that . . . (and here the dots are to be replaced by a T-theory). I am happy to accept this version, since it is equivalent to my own. (So far as I know, I never held the view he attributes to me which leaves unconnected the knowledge of what a theory of truth states and the knowledge that the theory is T-theoretical.)

Now let us consider the view which Foster thinks I should, and I know I do, hold. It cannot be said that on this view, knowledge of a language reduces to knowing how to translate it into another. The interpreter does, indeed, know that his knowledge consists in what is stated by a T-theory, a T-theory that is translational (satisfies Convention T). But there is no reason to suppose the interpreter can express his knowledge in any specific linguistic form, much less in any particular language.

Perhaps we should insist that a theory is a sentence or a set of sentences of some language. But to know a theory it is neither necessary nor sufficient to know that these sentences are true. Not sufficient since this could be known by someone who had no idea what the sentences meant, and not necessary since it is enough to know the truths the sentences of the theory express, and this does not require knowledge of the language of the theory.

Someone who can interpret English knows, for example, that an utterance of the sentence 'Snow is white' is true if and only if snow is white; he knows in addition that this fact is entailed by a translational theory—that it is not an accidental fact about that English sentence, but a fact that *interprets* the sentence. Once the point of putting things this way is clear, I see no harm in rephrasing what the interpreter knows in this case in a more familiar vein: he knows that 'Snow is white' in English *means that* snow is white.

It is clear, then, that my view does not make the ability to interpret a language depend on being able to translate that language into a familiar tongue. Perhaps it is worth reinforcing this point by tidying up a matter so far passed over. In natural languages indexical elements, like demonstratives and tense, mean that the truth conditions for many sentences must be made relative to the circumstances of their utterance. When this is done, the right side of the biconditional of a T-sentence never translates the sentence for which it is giving the truth conditions. In general, an adequate theory of truth uses no indexical devices, and so can contain no translations of a very large number and variety of sentences. With respect to these sentences, there is not even the illusion that interpretation depends on the ability to translate. (The 'means that' idiom does no better here.)

Foster thinks my grand plan is in ruins because in trying to harness the claim of T-theoreticity to secure interpretation I must use an intensional notion like the 'states' in 'The interpreter knows that some T-theory states that . . .'. But here he foists on me a goal I never had. My way of trying to give an account of language and meaning makes essential use of such concepts as those of belief and intention, and I do not believe it is possible to reduce these notions to anything more scientific or behaviouristic. What I have tried to do is give an account of meaning (interpretation) that makes no essential use of unexplained *linguistic* concepts. (Even this is a little stronger than what I think is possible.) It will ruin no plan of mine if in saying what an interpreter knows it is necessary to use a so-called intensional notion—one that consorts with belief and intention and the like.

Of course my project does require that all sentences of natural languages can be handled by a T-theory, and so if the intensional idioms resist such treatment, my plan has foundered. It seems to be the case, though the matter is not entirely simple or clear, that a theory of truth that satisfies anything like Convention T cannot allow an intensional semantics, and this has prompted me to try to show how an extensional semantics can handle what is special about belief sentences, indirect discourse, and other such sentences. Foster thinks my analysis will not do, but it is not easy to see how this is relevant to our debate. If his point is that *no* T-theory can give a satisfactory semantics for sentences that attribute attitudes, then all the discussion of how exactly to describe the competence of a speaker is simply irrelevant. But if some analysis is possible, mine or another, then what works for indirect discourse and sentences about belief and intention will presumably work also for the 'states' relation that worries Foster.

Foster is certainly right that the expression 'a T-theory states that' is what would usually be called a non-truth-functional sentential operator, since following it with materially equivalent sentences may produce results with divergent truth values. This leaves us with two problems (between which Foster does not perhaps sufficiently distinguish). The first is whether the paratactic analysis of indirect discourse can properly be applied in the present case; the other is whether my account of radical interpretation is threatened if the relevant notion of stating (whatever its semantics) conceals an unanalysed linguistic concept. The former problem is, I have just suggested, only marginally germane to our discussion; the latter is obviously central. I would, however, like to say something on both topics, since unless such an analysis is shown to be faulty I do propose a paratactic semantics for 'states that'.

The paratactic semantic approach to indirect discourse tells us to view an utterance of 'Galileo said that the Earth moves' as consisting of the utterance of two sentences, 'Galileo said that' and 'The Earth moves'. The 'that' refers to the second utterance, and the first utterance is true if and only if an utterance of Galileo's was the same in content as ('translates') the utterance to which the 'that' refers. (Foster wrongly says my analysis of 'Galileo said that' is 'Some utterance of Galileo and my last utterance make Galileo and me same-sayers'. This is not an analysis, but a rephrasal designed to give a reader a feeling for the semantics; an expository and heuristic device.)

Foster tries to prove my semantic analysis wrong by showing that it fails a translation test. This test requires that the translation of an utterance (as analysed) must state the same fact or proposition (I am using Foster's words) as the original utterance. Then he points out that a translation of 'Galileo said that' into French that preserved the reference of 'that' (on my analysis) would fail to convey to a French audience anything about the content of Galileo's remark. He takes this to show that on my analysis an ordinary utterance of a sentence like 'Galileo said that the Earth moves' fails to state what Galileo said; and so a parallel analysis of the 'states that' which my theory of interpretation needs will suffer from the same failure.

But what is this relation between utterances, of stating the same fact or proposition, that Foster has in mind? All he tells us is that reference must be preserved. This is surely not enough, however, since if any two utterances state the same fact when reference is preserved, it is very difficult to block a familiar proof that all true utterances state the same fact. The use of the word 'proposition' suggests that meaning must be preserved as well as reference. But if both reference and meaning must be preserved, it is easy to see that very few pairs of utterances can state the same fact provided the utterances contain indexical expressions. Leaving aside bilinguals, no French utterance can state any fact I do by using 'I', and I cannot twice state the same fact by saying 'I'm warm' twice. To judge my analysis wrong by these standards is simply to judge it wrong because it supposes that indirect discourse involves an indexical element. Failing further argument, the conclusion throws doubt on the standards, not the analysis.

Like Foster, I assume of course that a translator will render English indirect discourse into French in the usual way. In my view, he will do this by referring to a new utterance which he will have to supply. (The same thing goes on if I utter 'Galileo said that the Earth moves' twice.) To admit this is not 'to construe the paratactic version of oratio obliqua as a notational variant of the intensional version', as Foster urges. Notation has nothing to do with it; both the possible world semanticist and I accept the same notation. We differ on the semantic analysis. The point is that the translator is stating the same fact, not in Foster's sense, but in some more usual sense which often allows, or even requires, that translation change the reference when that reference is, to use Reichenbach's phrase, token reflexive.

Turning back to the prospects for a paratactic analysis of that troublesome 'states', we ought first to note that a slightly more appropriate word would be 'entails'. What we want is the semantics for utterances of sentences like 'Theory T entails that "Snow is white" is true in English if and only if snow is white'. And the claim must be that an utterance of this sentence is to be treated, for the purposes of semantic theory, as the utterance of two sentences, the first ending in a demonstrative which refers to the second utterance. Entailment (of this sort) is thus made out to be a relation between a theory and an utterance of the speaker who claims entailment. What is this relation? A reasonable suggestion is that it is the relative product of the relation of logical consequence between sentences and the relation of synonymy between sentences and utterances (perhaps of another language). If a theory T entails that 'Snow is white' is true in English if and only if snow is white, then T has as logical consequence a sentence synonomous with my utterance of ' "Snow is white" is true in English if and only if snow is white'.

Does not the second component bring in an appeal to a specifically linguistic notion—that of synonomy? Certainly: it is just the concept of translation we have been trying to elicit by placing conditions on a theory of truth. This does not make the account circular, for those conditions were stated, we have been assuming, in a non-question-begging way, without appeal to linguistic notions of the kind we want to explain. So the concept of synonomy or translation that lies concealed in the notion of entailment can be used without circularity when we come to set out what an interpreter knows. Indeed, in attributing to an interpreter the concept of a translational theory we have already made this assumption.

On a point of some importance, I think Foster is right. Even if everything I have said in defence of my formulation of what suffices for interpretation is right, it remains the case that nothing strictly constitutes a theory of meaning. A theory of truth, no matter how well selected, is not a theory of meaning, while the statement that a translational theory entails certain facts is not, because of the irreducible indexical elements in the sentences that express it, a theory in the formal sense. This does not, however, make it impossible to say what it is that an interpreter knows, and thus to give a satisfactory answer to one of the central problems of the philosophy of language.

LANGUAGE AND REALITY

13 *On the Very Idea of a Conceptual Scheme*

Philosophers of many persuasions are prone to talk of conceptual schemes. Conceptual schemes, we are told, are ways of organizing experience; they are systems of categories that give form to the data of sensation; they are points of view from which individuals, cultures, or periods survey the passing scene. There may be no translating from one scheme to another, in which case the beliefs, desires, hopes, and bits of knowledge that characterize one person have no true counterparts for the subscriber to another scheme. Reality itself is relative to a scheme: what counts as real in one system may not in another.

Even those thinkers who are certain there is only one conceptual scheme are in the sway of the scheme concept; even monotheists have religion. And when someone sets out to describe 'our conceptual scheme', his homey task assumes, if we take him literally, that there might be rival systems.

Conceptual relativism is a heady and exotic doctrine, or would be if we could make good sense of it. The trouble is, as so often in philosophy, it is hard to improve intelligibility while retaining the excitement. At any rate that is what I shall argue.

We are encouraged to imagine we understand massive conceptual change or profound contrasts by legitimate examples of a familiar sort. Sometimes an idea, like that of simultaneity as defined in relativity theory, is so important that with its addition a whole department of science takes on a new look. Sometimes revisions in the list of sentences held true in a discipline are so central that we may feel that the terms involved have changed their meanings. Languages that have evolved in distant times or places may differ extensively in their resources for dealing with one or another range of phenomena. What comes easily in one language may come hard in another, and this difference may echo significant dissimilarities in style and value.

But examples like these, impressive as they occasionally are, are not so extreme but that the changes and the contrasts can be explained and described using the equipment of a single language. Whorf, wanting to demonstrate that Hopi incorporates a metaphysics so alien to ours that Hopi and English cannot, as he puts it, 'be calibrated', uses English to convey the contents of sample Hopi sentences.[1] Kuhn is brilliant at saying what things were like before the revolution using—what else?—our post-revolutionary idiom.[2] Quine gives us a feel for the 'pre-individuative phase in the evolution of our conceptual scheme',[3] while Bergson tells us where we can go to get a view of a mountain undistorted by one or another provincial perspective.

The dominant metaphor of conceptual relativism, that of differing points of view, seems to betray an underlying paradox. Different points of view make sense, but only if there is a common co-ordinate system on which to plot them; yet the existence of a common system belies the claim of dramatic incomparability. What we need, it seems to me, is some idea of the considerations that set the limits to conceptual contrast. There are extreme suppositions that founder on paradox or contradiction; there are modest examples we have no trouble understanding. What determines where we cross from the merely strange or novel to the absurd?

We may accept the doctrine that associates having a language with having a conceptual scheme. The relation may be supposed to be this: where conceptual schemes differ, so do languages. But speakers of different languages may share a conceptual scheme provided there is a way of translating one language into the other. Studying the criteria of translation is therefore a way of focusing on criteria of identity for conceptual schemes. If conceptual schemes aren't associated with languages in this way, the original problem is needlessly doubled, for then we would have to imagine the mind, with its ordinary categories, operating with a language with *its* organizing structure. Under the circumstances we would certainly want to ask who is to be master.

[1] B. L. Whorf, 'The Punctual and Segmentative Aspects of Verbs in Hopi'.
[2] T. S. Kuhn, *The Structure of Scientific Revolutions.*
[3] W. V. Quine, 'Speaking of Objects', 24.

Alternatively, there is the idea that *any* language distorts reality, which implies that it is only wordlessly if at all that the mind comes to grips with things as they really are. This is to conceive language as an inert (though necessarily distorting) medium independent of the human agencies that employ it; a view of language that surely cannot be maintained. Yet if the mind can grapple without distortion with the real, the mind itself must be without categories and concepts. This featureless self is familiar from theories in quite different parts of the philosophical landscape. There are, for example, theories that make freedom consist in decisions taken apart from all desires, habits, and dispositions of the agent; and theories of knowledge that suggest that the mind can observe the totality of its own perceptions and ideas. In each case, the mind is divorced from the traits that constitute it; an inescapable conclusion from certain lines of reasoning, as I said, but one that should always persuade us to reject the premisses.

We may identify conceptual schemes with languages, then, or better, allowing for the possibility that more than one language may express the same scheme, sets of intertranslatable languages. Languages we will not think of as separable from souls; speaking a language is not a trait a man can lose while retaining the power of thought. So there is no chance that someone can take up a vantage point for comparing conceptual schemes by temporarily shedding his own. Can we then say that two people have different conceptual schemes if they speak languages that fail of intertranslatability?

In what follows I consider two kinds of case that might be expected to arise: complete, and partial, failures of translatability. There would be complete failure if no significant range of sentences in one language could be translated into the other; there would be partial failure if some range could be translated and some range could not (I shall neglect possible asymmetries.) My strategy will be to argue that we cannot make sense of total failure, and then to examine more briefly cases of partial failure.

First, then, the purported cases of complete failure. It is tempting to take a very short line indeed: nothing, it may be said, could count as evidence that some form of activity could not be interpreted in our language that was not at the same time evidence that that form of activity was not speech behaviour. If this were right, we probably ought to hold that a form of activity that cannot be interpreted as language in our language is not speech behaviour. Putting matters this way is unsatisfactory, however, for it comes to little more than making translatability into a familiar tongue a criterion of languagehood. As fiat, the thesis lacks the appeal of self-evidence; if it is a truth, as I think it is, it should emerge as the conclusion of an argument.

The credibility of the position is improved by reflection on the close relations between language and the attribution of attitudes such as belief, desire, and intention. On the one hand, it is clear that speech requires a multitude of finely discriminated intentions and beliefs. A person who asserts that perseverance keeps honour bright must, for example, represent himself as believing that perseverance keeps honour bright, and he must intend to represent himself as believing it. On the other hand, it seems unlikely that we can intelligibly attribute attitudes as complex as these to a speaker unless we can translate his words into ours. There can be no doubt that the relation between being able to translate someone's language and being able to describe his attitudes is very close. Still, until we can say more about *what* this relation is, the case against untranslatable languages remains obscure.

It is sometimes thought that translatability into a familiar language, say English, cannot be a criterion of languagehood on the grounds that the relation of translatability is not transitive. The idea is that some language, say Saturnian, may be translatable into English, and some further language, like Plutonian, may be translatable into Saturnian, while Plutonian is not translatable into English. Enough translatable differences may add up to an untranslatable one. By imagining a sequence of languages, each close enough to the one before to be acceptably translated into it, we can imagine a language so different from English as to resist totally translation into it. Corresponding to this distant language would be a system of concepts altogether alien to us.

This exercise does not, I think, introduce any new element into the discussion. For we should have to ask how we recognized that what the Saturnian was doing was *translating* Plutonian (or anything else). The Saturnian speaker might tell us that that was what he was doing or rather we might for a moment assume that that was what he was telling us. But then it would occur to us to wonder whether our translations of Saturnian were correct.

According to Kuhn, scientists operating in different scientific traditions (within different 'paradigms') 'work in different worlds'.[4] Strawson's *The Bounds of Sense* begins with the remark that 'It is possible to imagine kinds of worlds very different from the world as we know it'.[5] Since there is at most one world, these pluralities are metaphorical or merely imagined. The metaphors are, however, not at all the same. Strawson invites us to imagine possible non-actual worlds, worlds that might be described, using our present language, by redistributing truth values over sentences in various systematic ways. The clarity of the contrasts between worlds in this case depends on supposing our scheme of concepts, our descriptive resources, to remain fixed. Kuhn, on the other hand, wants us to think of different observers of the same world who come to it with incommensurable systems of concepts. Strawson's many imagined worlds are seen or heard or described from the same point of view; Kuhn's one world is seen from different points of view. It is the second metaphor we want to work on.

The first metaphor requires a distinction within language of concept and content: using a fixed system of concepts (words with fixed meanings) we describe alternative universes. Some sentences will be true simply because of the concepts or meanings involved, others because of the way of the world. In describing possible worlds, we play with sentences of the second kind only.

The second metaphor suggests instead a dualism of quite a different sort, a dualism of total scheme (or language) and uninterpreted content. Adherence to the second dualism, while not inconsistent with adherence to the first, may be encouraged by attacks on the first. Here is how it may work.

To give up the analytic-synthetic distinction as basic to the understanding of language is to give up the idea that we can clearly distinguish between theory and language. Meaning, as we might loosely use the word, is contaminated by theory, by what is held to be true. Feyerabend puts it this way:

> Our argument against meaning invariance is simple and clear. It proceeds from the fact that usually some of the principles involved in the determinations of the meanings of older theories or points of views are inconsistent with the new . . . theories. It points out that it is natural to resolve this contradiction by eliminating the troublesome . . . older principles, and to replace them by principles, or theorems, of a new . . . theory. And it

[4] T. S. Kuhn, *The Structure of Scientific Revolutions*, 134.
[5] P. Strawson, *The Bounds of Sense*, 15.

concludes by showing that such a procedure will also lead to the elimination of the old meanings.[6]

We may now seem to have a formula for generating distinct conceptual schemes. We get a new out of an old scheme when the speakers of a language come to accept as true an important range of sentences they previously took to be false (and, of course, vice versa). We must not describe this change simply as a matter of their coming to view old falsehoods as truths, for a truth is a proposition, and what they come to accept, in accepting a sentence as true, is not the same thing that they rejected when formerly they held the sentence to be false. A change has come over the meaning of the sentence because it now belongs to a new language.

This picture of how new (perhaps better) schemes result from new and better science is very much the picture philosophers of science, like Putnam and Feyerabend, and historians of science, like Kuhn, have painted for us. A related idea emerges in the suggestion of some other philosophers, that we could improve our conceptual lot if we were to tune our language to an improved science. Thus both Quine and Smart, in somewhat different ways, regretfully admit that our present ways of talking make a serious science of behaviour impossible. (Wittgenstein and Ryle have said similar things without regret.) The cure, Quine and Smart think, is to change how we talk. Smart advocates (and predicts) the change in order to put us on the scientifically straight path of materialism: Quine is more concerned to clear the way for a purely extensional language. (Perhaps I should add that I think our actual scheme and language are best understood as extensional and materialist.)

If we were to follow this advice, I do not myself think science or understanding would be advanced, though possibly morals would. But the present question is only whether, if such changes were to take place, we should be justified in calling them alterations in the basic conceptual apparatus. The difficulty in so calling them is easy to appreciate. Suppose that in my office of Minister of Scientific Language I want the new man to stop using words that refer, say, to emotions, feelings, thoughts, and intentions, and to talk instead of the physiological states and happenings that are assumed to be more or less identical with the mental riff and raff. How do I tell whether my advice has been heeded if the new man speaks a new language?

[6] P. Feyerabend, 'Explanation, Reduction, and Empiricism', 82.

For all I know, the shiny new phrases, though stolen from the old language in which they refer to physiological stirrings, may in his mouth play the role of the messy old mental concepts.

The key phrase is: for all I know. What is clear is that retention of some or all of the old vocabulary in itself provides no basis for judging the new scheme to be the same as, or different from, the old. So what sounded at first like a thrilling discovery—that truth is relative to a conceptual scheme—has not so far been shown to be anything more than the pedestrian and familiar fact that the truth of a sentence is relative to (among other things) the language to which it belongs. Instead of living in different worlds, Kuhn's scientists may, like those who need Webster's dictionary, be only words apart.

Giving up the analytic-synthetic distinction has not proven a help in making sense of conceptual relativism. The analytic-synthetic distinction is however explained in terms of something that may serve to buttress conceptual relativism, namely the idea of empirical content. The dualism of the synthetic and the analytic is a dualism of sentences some of which are true (or false) both because of what they mean and because of their empirical content, while others are true (or false) by virtue of meaning alone, having no empirical content. If we give up the dualism, we abandon the conception of meaning that goes with it, but we do not have to abandon the idea of empirical content: we can hold, if we want, that *all* sentences have empirical content. Empirical content is in turn explained by reference to the facts, the world, experience, sensation, the totality of sensory stimuli, or something similar. Meanings gave us a way to talk about categories, the organizing structure of language, and so on; but it is possible, as we have seen, to give up meanings and analyticity while retaining the idea of language as embodying a conceptual scheme. Thus in place of the dualism of the analytic-synthetic we get the dualism of conceptual scheme and empirical content. The new dualism is the foundation of an empiricism shorn of the untenable dogmas of the analytic-synthetic distinction and reductionism—shorn, that is, of the unworkable idea that we can uniquely allocate empirical content sentence by sentence.

I want to urge that this second dualism of scheme and content, of organizing system and something waiting to be organized, cannot be made intelligible and defensible. It is itself a dogma of empiricism, the third dogma. The third, and perhaps the last, for if we give it up it is not clear that there is anything distinctive left to call empiricism.

The scheme-content dualism has been formulated in many ways. Here are some examples. The first comes from Whorf, elaborating on a theme of Sapir's. Whorf says that:

> ... language produces an organization of experience. We are inclined to think of language simply as a technique of expression, and not to realize that language first of all is a classification and arrangement of the stream of sensory experience which results in a certain world-order ... In other words, language does in a cruder but also in a broader and more versatile way the same thing that science does ... We are thus introduced to a new principle of relativity, which holds that all observers are not led by the same physical evidence to the same picture of the universe, unless their linguistic backgrounds are similar, or can in some way be calibrated.[7]

Here we have all the required elements: language as the organizing force, not to be distinguished clearly from science; what is organized, referred to variously as 'experience', 'the stream of sensory experience', and 'physical evidence'; and finally, the failure of intertranslatability ('calibration'). The failure of intertranslatability is a necessary condition for difference of conceptual schemes; the common relation to experience or the evidence is what is supposed to help us make sense of the claim that it is languages or schemes that are under consideration when translation fails. It is essential to this idea that there be something neutral and common that lies outside all schemes. This common something cannot, of course, be the *subject matter* of contrasting languages, or translation would be possible. Thus Kuhn has recently written:

> Philosophers have now abandoned hope of finding a pure sense-datum language ... but many of them continue to assume that theories can be compared by recourse to a basic vocabulary consisting entirely of words which are attached to nature in ways that are unproblematic and, to the extent necessary, independent of theory ... Feyerabend and I have argued at length that no such vocabulary is available. In the transition from one theory to the next words change their meanings or conditions of applicability in subtle ways. Though most of the same signs are used before and after a revolution—e.g. force, mass, element, compound, cell—the way in which some of them attach to nature has somehow changed. Successive theories are thus, we say, incommensurable.[8]

'Incommensurable' is, of course, Kuhn and Feyerabend's word for 'not intertranslatable'. The neutral content waiting to be organized is supplied by nature.

[7] B. L. Whorf, 'The Punctual and Segmentative Aspects of Verbs in Hopi', 55.
[8] T. S. Kuhn, 'Reflections on my Critics', 266, 267.

Feyerabend himself suggests that we may compare contrasting schemes by 'choosing a point of view outside the system or the language'. He hopes we can do this because 'there is still human experience as an actually existing process'[9] independent of all schemes.

The same, or similar, thoughts are expressed by Quine in many passages: 'The totality of our so-called knowledge or beliefs . . . is a man-made fabric which impinges on experience only along the edges . . .';[10] '. . . total science is like a field of force whose boundary conditions are experience';[11] 'As an empiricist I . . . think of the conceptual scheme of science as a tool . . . for predicting future experience in the light of past experience.'[12] And again:

> We persist in breaking reality down somehow into a multiplicity of identifiable and discriminable objects . . . We talk so inveterately of objects that to say we do so seems almost to say nothing at all; for how else is there to talk? It is hard to say how else there is to talk, not because our objectifying pattern is an invariable trait of human nature, but because we are bound to adapt any alien pattern to our own in the very process of understanding or translating the alien sentences.[13]

The test of difference remains failure or difficulty of translation: '. . . to speak of that remote medium as radically different from ours is to say no more than that the translations do not come smoothly.'[14] Yet the roughness may be so great that the alien has an 'as yet unimagined pattern beyond individuation'.[15]

The idea is then that something is a language, and associated with a conceptual scheme, whether we can translate it or not, if it stands in a certain relation (predicting, organizing, facing, or fitting) experience (nature, reality, sensory promptings). The problem is to say what the relation is, and to be clearer about the entities related.

The images and metaphors fall into two main groups: conceptual schemes (languages) either *organize* something, or they *fit* it (as in 'he warps his scientific heritage to fit his . . . sensory promptings'[16]). The first group contains also *systematize*, *divide up* (the stream of experience); further examples of the second group are *predict*, *account for*, *face* (the tribunal of experience). As for the entities that

[9] P. Feyerabend, 'Problems of Empiricism', 214.
[10] W. V. Quine, 'Two Dogmas of Empiricism', 42.
[11] Ibid.
[12] Ibid., 44.
[13] W. V. Quine, 'Speaking of Objects', 1.
[14] Ibid., 25.
[15] Ibid., 24.
[16] W. V. Quine, 'Two Dogmas of Empiricism', 46.

get organized, or which the scheme must fit, I think again we may detect two main ideas: either it is reality (the universe, the world, nature), or it is experience (the passing show, surface irritations, sensory promptings, sense-data, the given).

We cannot attach a clear meaning to the notion of organizing a single object (the world, nature etc.) unless that object is understood to contain or consist in other objects. Someone who sets out to organize a closet arranges the things in it. If you are told not to organize the shoes and shirts, but the closet itself, you would be bewildered. How would you organize the Pacific Ocean? Straighten out its shores, perhaps, or relocate its islands, or destroy its fish.

A language may contain simple predicates whose extensions are matched by no simple predicates, or even by any predicates at all, in some other language. What enables us to make this point in particular cases is an ontology common to the two languages, with concepts that individuate the same objects. We can be clear about breakdowns in translation when they are local enough, for a background of generally successful translation provides what is needed to make the failures intelligible. But we were after larger game: we wanted to make sense of there being a language we could not translate at all. Or, to put the point differently, we were looking for a criterion of languagehood that did not depend on, or entail, translatability into a familiar idiom. I suggest that the image of organizing the closet of nature will not supply such a criterion.

How about the other kind of object, experience? Can we think of a language organizing *it*? Much the same difficulties recur. The notion of organization applies only to pluralities. But whatever plurality we take experience to consist in—events like losing a button or stubbing a toe, having a sensation of warmth or hearing an oboe—we will have to individuate according to familiar principles. A language that organizes *such* entities must be a language very like our own.

Experience (and its classmates like surface irritations, sensations, and sense-data) also makes another and more obvious trouble for the organizing idea. For how could something count as a language that organized *only* experiences, sensations, surface irritations, or sense-data? Surely knives and forks, railroads and mountains, cabbages and kingdoms also need organizing.

This last remark will no doubt sound inappropriate as a response to the claim that a conceptual scheme is a way of coping with sensory experience; and I agree that it is. But what was under consideration was the idea of *organizing* experience, not the idea of *coping with* (or fitting or facing) experience. The reply was apropos of the former, not the latter, concept. So now let's see whether we can do better with the second idea.

When we turn from talk of organization to talk of fitting we turn our attention from the referential apparatus of language—predicates, quantifiers, variables, and singular terms—to whole sentences. It is sentences that predict (or are used to predict), sentences that cope or deal with things, that fit our sensory promptings, that can be compared or confronted with the evidence. It is sentences also that face the tribunal of experience, though of course they must face it together.

The proposal is not that experiences, sense-data, surface irritations, or sensory promptings are the sole subject matter of language. There is, it is true, the theory that talk about brick houses on Elm Street is ultimately to be construed as being about sense data or perceptions, but such reductionistic views are only extreme, and implausible, versions of the general position we are considering. The general position is that sensory experience provides all the *evidence* for the acceptance of sentences (where sentences may include whole theories). A sentence or theory fits our sensory promptings, successfully faces the tribunal of experience, predicts future experience, or copes with the pattern of our surface irritations, provided it is borne out by the evidence.

In the common course of affairs, a theory may be borne out by the available evidence and yet be false. But what is in view here is not just actually available evidence; it is the totality of possible sensory evidence past, present, and future. We do not need to pause to contemplate what this might mean. The point is that for a theory to fit or face up to the totality of possible sensory evidence is for that theory to be true. If a theory quantifies over physical objects, numbers, or sets, what it says about these entities is true provided the theory as a whole fits the sensory evidence. One can see how, from this point of view, such entities might be called posits. It is reasonable to call something a posit if it can be contrasted with something that is not. Here the something that is not is sensory experience—at least that is the idea.

The trouble is that the notion of fitting the totality of experience, like the notion of fitting the facts, or of being true to the facts, adds nothing intelligible to the simple concept of being true. To speak of sensory experience rather than the evidence, or just the facts, expresses a view about the source or nature of evidence, but it does not add a new entity to the universe against which to test conceptual schemes. The totality of sensory evidence is what we want provided it is all the evidence there is; and all the evidence there is is just what it takes to make our sentences or theories true. Nothing, however, no *thing*, makes sentences and theories true: not experience, not surface irritations, not the world, can make a sentence true. *That* experience takes a certain course, that our skin is warmed or punctured, that the universe is finite, these facts, if we like to talk that way, make sentences and theories true. But this point is put better without mention of facts. The sentence 'My skin is warm' is true if and only if my skin is warm. Here there is no reference to a fact, a world, an experience, or a piece of evidence.[17]

Our attempt to characterize languages or conceptual schemes in terms of the notion of fitting some entity has come down, then, to the simple thought that something is an acceptable conceptual scheme or theory if it is true. Perhaps we better say *largely* true in order to allow sharers of a scheme to differ on details. And the criterion of a conceptual scheme different from our own now becomes: largely true but not translatable. The question whether this is a useful criterion is just the question how well we understand the notion of truth, as applied to language, independent of the notion of translation. The answer is, I think, that we do not understand it independently at all.

We recognize sentences like ' "Snow is white" is true if and only if snow is white' to be trivially true. Yet the totality of such English sentences uniquely determines the extension of the concept of truth for English. Tarski generalized this observation and made it a test of theories of truth: according to Tarski's Convention T, a satisfactory theory of truth for a language L must entail, for every sentence *s* of L, a theorem of the form '*s* is true if and only if *p*' where '*s*' is replaced by a description of *s* and '*p*' by *s* itself if L is English, and by a translation of *s* into English if L is not English.[18] This isn't, of course, a definition of truth, and it doesn't hint that there is a single definition or theory that applies to languages generally. Nevertheless, Convention T suggests, though it cannot state, an important feature

[17] See Essay 3.
[18] A. Tarski, 'The Concept of Truth in Formalized Languages'.

common to all the specialized concepts of truth. It succeeds in doing this by making essential use of the notion of translation into a language we know. Since Convention T embodies our best intuition as to how the concept of truth is used, there does not seem to be much hope for a test that a conceptual scheme is radically different from ours if that test depends on the assumption that we can divorce the notion of truth from that of translation.

Neither a fixed stock of meanings, nor a theory-neutral reality, can provide, then, a ground for comparison of conceptual schemes. It would be a mistake to look further for such a ground if by that we mean something conceived as common to incommensurable schemes. In abandoning this search, we abandon the attempt to make sense of the metaphor of a single space within which each scheme has a position and provides a point of view.

I turn now to the more modest approach: the idea of partial rather than total failure of translation. This introduces the possibility of making changes and contrasts in conceptual schemes intelligible by reference to the common part. What we need is a theory of translation or interpretation that makes no assumptions about shared meanings, concepts, or beliefs.

The interdependence of belief and meaning springs from the interdependence of two aspects of the interpretation of speech behaviour: the attribution of beliefs and the interpretation of sentences. We remarked before that we can afford to associate conceptual schemes with languages because of these dependencies. Now we can put the point in a somewhat sharper way. Allow that a man's speech cannot be interpreted except by someone who knows a good deal about what the speaker believes (and intends and wants), and that fine distinctions between beliefs are impossible without understood speech; how then are we to interpret speech or intelligibly to attribute beliefs and other attitudes? Clearly we must have a theory that simultaneously accounts for attitudes and interprets speech, and which assumes neither.

I suggest, following Quine, that we may without circularity or unwarranted assumptions accept certain very general attitudes towards sentences as the basic evidence for a theory of radical interpretation. For the sake of the present discussion at least we may depend on the attitude of accepting as true, directed to sentences, as the crucial notion. (A more full-blooded theory would look to other attitudes towards sentences as well, such as wishing true, wondering whether true, intending to make true, and so on.) Attitudes are indeed involved here, but the fact that the main issue is not begged can be seen from this: if we merely know that someone holds a certain sentence to be true, we know neither what he means by the sentence nor what belief his holding it true represents. His holding the sentence true is thus the vector of two forces: the problem of interpretation is to abstract from the evidence a workable theory of meaning and an acceptable theory of belief.

The way this problem is solved is best appreciated from undramatic examples. If you see a ketch sailing by and your companion says, 'Look at that handsome yawl', you may be faced with a problem of interpretation. One natural possibility is that your friend has mistaken a ketch for a yawl, and has formed a false belief. But if his vision is good and his line of sight favourable it is even more plausible that he does not use the word 'yawl' quite as you do, and has made no mistake at all about the position of the jigger on the passing yacht. We do this sort of off the cuff interpretation all the time, deciding in favour of reinterpretation of words in order to preserve a reasonable theory of belief. As philosophers we are peculiarly tolerant of systematic malapropism, and practised at interpreting the result. The process is that of constructing a viable theory of belief and meaning from sentences held true.

Such examples emphasize the interpretation of anomalous details against a background of common beliefs and a going method of translation. But the principles involved must be the same in less trivial cases. What matters is this: if all we know is what sentences a speaker holds true, and we cannot assume that his language is our own, then we cannot take even a first step towards interpretation without knowing or assuming a great deal about the speaker's beliefs. Since knowledge of beliefs comes only with the ability to interpret words, the only possibility at the start is to assume general agreement on beliefs. We get a first approximation to a finished theory by assigning to sentences of a speaker conditions of truth that actually obtain (in our own opinion) just when the speaker holds those sentences true. The guiding policy is to do this as far as possible, subject to considerations of simplicity, hunches about the effects of social conditioning, and of course our common-sense, or scientific, knowledge of explicable error.

The method is not designed to eliminate disagreement, nor can it; its purpose is to make meaningful disagreement possible, and this depends entirely on a foundation—*some* foundation—in agreement. The agreement may take the form of widespread sharing of sentences held true by speakers of 'the same language', or agreement in the large mediated by a theory of truth contrived by an interpreter for speakers of another language.

Since charity is not an option, but a condition of having a workable theory, it is meaningless to suggest that we might fall into massive error by endorsing it. Until we have successfully established a systematic correlation of sentences held true with sentences held true, there are no mistakes to make. Charity is forced on us; whether we like it or not, if we want to understand others, we must count them right in most matters. If we can produce a theory that reconciles charity and the formal conditions for a theory, we have done all that could be done to ensure communication. Nothing more is possible, and nothing more is needed.

We make maximum sense of the words and thoughts of others when we interpret in a way that optimizes agreement (this includes room, as we said, for explicable error, i.e. differences of opinion). Where does this leave the case for conceptual relativism? The answer is, I think, that we must say much the same thing about differences in conceptual scheme as we say about differences in belief: we improve the clarity and bite of declarations of difference, whether of scheme or opinion, by enlarging the basis of shared (translatable) language or of shared opinion. Indeed, no clear line between the cases can be made out. If we choose to translate some alien sentence rejected by its speakers by a sentence to which we are strongly attached on a community basis, we may be tempted to call this a difference in schemes; if we decide to accommodate the evidence in other ways, it may be more natural to speak of a difference of opinion. But when others think differently from us, no general principle, or appeal to evidence, can force us to decide that the difference lies in our beliefs rather than in our concepts.

We must conclude, I think, that the attempt to give a solid meaning to the idea of conceptual relativism, and hence to the idea of a conceptual scheme, fares no better when based on partial failure of translation than when based on total failure. Given the underlying methodology of interpretation, we could not be in a position to judge that others had concepts or beliefs radically different from our own.

It would be wrong to summarize by saying we have shown how

communication is possible between people who have different schemes, a way that works without need of what there cannot be, namely a neutral ground, or a common co-ordinate system. For we have found no intelligible basis on which it can be said that schemes are different. It would be equally wrong to announce the glorious news that all mankind—all speakers of language, at least—share a common scheme and ontology. For if we cannot intelligibly say that schemes are different, neither can we intelligibly say that they are one.

In giving up dependence on the concept of an uninterpreted reality, something outside all schemes and science, we do not relinquish the notion of objective truth—quite the contrary. Given the dogma of a dualism of scheme and reality, we get conceptual relativity, and truth relative to a scheme. Without the dogma, this kind of relativity goes by the board. Of course truth of sentences remains relative to language, but that is as objective as can be. In giving up the dualism of scheme and world, we do not give up the world, but re-establish unmediated touch with the familiar objects whose antics make our sentences and opinions true or false.

14 *The Method of Truth in Metaphysics*

In sharing a language, in whatever sense this is required for communication, we share a picture of the world that must, in its large features, be true. It follows that in making manifest the large features of our language, we make manifest the large features of reality. One way of pursuing metaphysics is therefore to study the general structure of our language. This is not, of course, the sole true method of metaphysics; there is no such. But it is one method, and it has been practised by philosophers as widely separated by time or doctrine as Plato, Aristotle, Hume, Kant, Russell, Frege, Wittgenstein, Carnap, Quine, and Strawson. These philosophers have not, it goes without saying, agreed on what the large features of language are, or on how they may best be studied and described; the metaphysical conclusions have in consequence been various.

The method I will describe and recommend is not new; every important feature of the method can be found in one philosopher or another, and the leading idea is implicit in much of the best work in philosophy of language. What is new is the explicit formulation of the approach, and the argument for its philosophical importance. I begin with the argument; then comes a description of the method: finally, some applications are sketched.

I

Why must our language—any language—incorporate or depend upon a largely correct, shared, view of how things are? First consider why those who can understand one another's speech must share a view of the world, whether or not that view is correct. The reason is that we damage the intelligibility of our readings of the utterances of others when our method of reading puts others into what we take to be broad error. We can make sense of differences all right, but only against a background of shared belief. What is shared does not in general call for comment; it is too dull, trite, or familiar to stand notice. But without a vast common ground, there is no place for disputants to have their quarrel. Of course, we can no more agree than disagree with someone else without much mutuality; but perhaps this is obvious.

Beliefs are identified and described only within a dense pattern of beliefs. I can believe a cloud is passing before the sun, but only because I believe there is a sun, that clouds are made of water vapour, that water can exist in liquid or gaseous form; and so on, without end. No particular list of further beliefs is required to give substance to my belief that a cloud is passing before the sun; but some appropriate set of related beliefs must be there. If I suppose that you believe a cloud is passing before the sun, I suppose you have the right sort of pattern of beliefs to support that one belief, and these beliefs I assume you to have must, to do their supporting work, be enough like my beliefs to justify the description of your belief as a belief that a cloud is passing before the sun. If I am right in attributing the belief to you, then you must have a pattern of beliefs much like mine. No wonder, then, I can interpret your words correctly only by interpreting so as to put us largely in agreement.

It may seem that the argument so far shows only that good interpretation breeds concurrence, while leaving quite open the question whether what is agreed upon is true. And certainly agreement, no matter how widespread, does not guarantee truth. This observation misses the point of the argument, however. The basic claim is that much community of belief is needed to provide a basis for communication or understanding; the extended claim should then be that objective error can occur only in a setting of largely true belief. Agreement does not make for truth, but much of what is agreed must be true if some of what is agreed is false.

Just as too much attributed error risks depriving the subject of his subject matter, so too much actual error robs a person of things to go wrong about. When we want to interpret, we work on one or another assumption about the general pattern of agreement. We suppose that much of what we take to be common is true, but we cannot, of course, assume we know where the truth lies. We cannot interpret on the basis of known truths, not because we know none, but because we do not always know which they are. We do not need to be omniscient to interpret, but there is nothing absurd in the idea of an omniscient interpreter; he attributes beliefs to others, and interprets their speech on the basis of his own beliefs, just as the rest of us do. Since he does this as the rest of us do, he perforce finds as much agreement as is needed to make sense of his attributions and interpretations; and in this case, of course, what is agreed is by hypothesis true. But now it is plain why massive error about the world is simply unintelligible, for to suppose it intelligible is to suppose there could be an interpreter (the omniscient one) who correctly interpreted someone else as being massively mistaken, and this we have shown to be impossible.

II

Successful communication proves the existence of a shared, and largely true, view of the world. But what led us to demand the common view was the recognition that sentences held true—the linguistic representatives of belief—determine the meanings of the words they contain. Thus the common view shapes the shared language. This is why it is plausible to hold that by studying the most general aspects of language we will be studying the most general aspects of reality. It remains to say how these aspects may be identified and described.

Language is an instrument of communication because of its semantic dimension, the potentiality for truth or falsehood of its sentences, or better, of its utterances and inscriptions. The study of what sentences are true is in general the work of the various sciences; but the study of truth conditions is the province of semantics. What we must attend to in language, if we want to bring into relief general features of the world, is what it is in general for a sentence in the language to be true. The suggestion is that if the truth conditions of sentences are placed in the context of a comprehensive theory, the linguistic structure that emerges will reflect large features of reality.

The aim is a theory of truth for a reasonably powerful and significant part of a natural language. The scope of the theory—how much of the language is captured by the theory, and how convincingly—will be one factor on which the interest of any metaphysical results depends. The theory must show us how we can view each of a potential infinity of sentences as composed from a finite stock of semantically significant atoms (roughly, words) by means of a finite number of applications of a finite number of rules of composition. It must then give the truth conditions of each sentence (relative to the circumstances of its utterance) on the basis of its composition. The theory may thus be said to explain the conditions of truth of an utterance of a sentence on the basis of the roles of the words in the sentence.

Much here is owed to Frege. Frege saw the importance of giving an account of how the truth of a sentence depends on the semantic features of its parts, and he suggested how such an account could be given for impressive stretches of natural language. His method was one now familiar: he introduced a standardized notation whose syntax directly reflected the intended interpretation, and then urged that the new notation, as interpreted, had the same expressive power as important parts of natural language. Or rather, not quite the same expressive power, since Frege believed natural language was defective in some respects, and he regarded his new language as an improvement.

Frege was concerned with the semantic structure of sentences, and with semantic relations between sentences, in so far as these generated entailments. But he cannot be said to have conceived the idea of a comprehensive formal theory of truth for a language as a whole. One consequence was a lack of interest in the semantic paradoxes. Another was an apparent willingness to accept an infinity of meanings (senses) and referents for every denoting phrase in the language.

Because Frege took the application of function to argument to be the sole mode of semantic combination, he was bound to treat sentences as a kind of name—the name of a truth value. Seen simply as an artful dodge on the way to characterizing the truth conditions of sentences, this device of Frege's is unexceptionable. But since sentences do not operate in language the way names do, Frege's approach undermines confidence that the ontology he needs to work his semantics has any direct connection with the ontology implicit in natural language. It is not clear, then, what one can learn about metaphysics from Frege's method. (I certainly do not mean by this that we can't learn about metaphysics from Frege's work; but to see how, arguments different from mine must be marshalled.)

Quine provided an essential ingredient for the project at hand by showing how a holistic approach to the problem of understanding a language supplies the needed empirical foundation. If metaphysical conclusions are to be drawn from a theory of truth in the way that I propose, the approach to language must be holistic. Quine himself does not see holism as having such direct metaphysical significance, however, and for a number of reasons. First, Quine has not made the theory of truth central either as a key to the ontology of a language, or as a test of logical form. Second, like Frege, he views a satisfactorily regimented language as an improvement on natural language rather than as part of a theory about it. In one important respect, Quine seems even to go beyond Frege, for where Frege thinks his notation makes for better language, Quine thinks it also makes for better science. As a consequence, Quine ties his metaphysics to his canonical notation rather than to natural language; as he puts it, 'The quest of a simplest, clearest overall pattern of canonical notation is not to be distinguished from a quest of ultimate categories, a limning of the most general traits of reality.'[1]

The formal languages towards which I gravitate—first-order languages with standard logic—are those preferred by Quine, But our reasons for this choice diverge somewhat. Such languages please Quine because their logic is simple, and the scientifically respectable parts of natural language can be translated into them; and with this I agree. But since I am not interested in improving on natural language, but in understanding it, I view formal languages or canonical notations as devices for exploring the structure of natural language. We know how to give a theory of truth for the formal language; so if we also knew how to transform the sentences of a natural language systematically into sentences of the formal language, we would have a theory of truth for the natural language. From this point of view, standard formal languages are intermediate devices to assist us in treating natural languages as more complex formal languages.

Tarski's work on truth definitions for formalized languages serves as inspiration for the kind of theory of truth that is wanted for natural languages.[2] The method works by enumerating the semantic properties of the items in a finite vocabulary, and on this basis recursively characterizes truth for each of the infinity of sentences.

[1] W. V. Quine, *Word and Object*, 161.
[2] A. Tarski, 'The Concept of Truth in Formalized Languages'.

Truth is reached from the basis by the intervention of a subtle and powerful concept (satisfaction) which relates both sentences and non-sentential expressions to objects in the world. An important feature of Tarski's approach is that a characterization of a truth predicate '*x* is true in L' is accepted only if it entails, for each sentence of the language L, a theorem of the form '*x* is true in L if and only if . . .' with '*x*' replaced by a description of the sentence and the dots replaced by a translation of the sentence into the language of the theory.

It is evident that these theorems, which we may call T-sentences, require a predicate that holds of just the true sentences of L. It is also plain, from the fact that the truth conditions for a sentence translate that sentence (i.e., what appears to the right of the 'if and only if' in a T-sentence translates the sentence described on the left), that the theory shows how to characterize truth for any given sentence without appeal to conceptual resources not available in that sentence.

These remarks are only roughly correct. A theory of truth for a natural language must relativize the truth of a sentence to the circumstances of utterance, and when this is done the truth conditions given by a T-sentence will no longer translate the described sentence, nor will it be possible to avoid using concepts that are, perhaps, semantical, in giving the truth conditions of sentences with indexical elements. More important, the notion of translation, which can be made precise for artificial languages on which interpretations are imposed by fiat, has no precise or even clear application to natural languages.

For these, and other reasons, it is important to stress that a theory of truth for a natural language (as I conceive it) differs widely in both aim and interest from Tarski's truth definitions. Sharpness of application is lost, and with it most of what concerns mathematicians and logicians: consequences for consistency, for example. Tarski could take translation as syntactically specified, and go on to define truth. But in application to a natural language it makes more sense to assume a partial understanding of truth, and use the theory to throw light on meaning, interpretation, and translation.[3] Satisfaction of Tarski's Convention T remains a desideratum of a theory, but is no longer available as a formal test.

[3] See Essays 9 and 10.

What a theory of truth does for a natural language is reveal structure. In treating each sentence as composed in accountable ways out of a finite number of truth-relevant words, it articulates this structure. When we study terms and sentences directly, not in the light of a comprehensive theory, we must bring metaphysics to language; we assign roles to words and sentences in accord with the categories we independently posit on epistemological or metaphysical grounds. Operating in this way, philosophers ponder such questions as whether there must be entities, perhaps universals, that correspond to predicates, or non-existent entities to correspond to non-denoting names or descriptions; or they argue that sentences do, or do not, correspond to facts or propositions.

A different light is shed on these matters when we look for a comprehensive theory of truth, for such a theory makes its own unavoidable demands.

III

Now let us consider some applications. We noticed that the requirement that the truth conditions of a sentence be given using only the conceptual resources of that sentence is not entirely clear where it seems that it can be met, nor everywhere applicable. The cases that invite exception are sentences that involve demonstratives, and here the cure of the difficulty is relatively simple.[4] These cases aside, the requirement, for all its obscurity, has, I think, important implications.

Suppose we were to admit a rule like this as part of a theory of truth: 'A sentence consisting of a singular term followed by a one-place predicate is true if and only if the object named by the singular term belongs to the class determined by the predicate.'[5] This rule offends the requirement, for if the rule were admitted, the T-sentence for 'Socrates is wise' would be ' "Socrates is wise" is true if and only if the object named by "Socrates" belongs to the class determined by the predicate "is wise",' and here the statement of truth conditions involves two semantic concepts (naming and determining a class) not plausibly among the conceptual resources of 'Socrates is wise'.

It would be easy to get from the tendentious T-sentence just

[4] See S. Weinstein, 'Truth and Demonstratives'.

[5] Compare R. Carnap, *Meaning and Necessity*, 5.

mentioned to the non-committal and admissible '"Socrates is wise" is true if and only if Socrates is wise' if the theory also contained as postulates statements that the object named by 'Socrates' is Socrates and that x belongs to the class determined by the predicate 'is wise' if and only if x is wise. If enough such postulates are available to care for all proper names and primitive predicates, the results are clear. First, T-sentences free from unwanted semantic terms would be available for all the sentences involved; and the extra semantic terms would be unnecessary. For there would have to be a postulate for each name and predicate, and this there could be only if the list of names and primitive predicates were finite. But if the list were finite, there would be only a finite number of sentences consisting of a name and a one-place predicate, and nothing would stand in the way of giving the truth conditions for all such sentences straight off—the T-sentences themselves could serve as the axioms.

The example illustrates how keeping the vocabulary finite may allow the elimination of semantic concepts; it also shows how the demand for a satisfactory theory has ontological consequences. Here, the call for entities to correspond to predicates disappears when the theory is made to produce T-sentences without excess semantic baggage. Indeed in the case at hand the theory does not need to put expressions and objects into explicit correspondence at all, and so no ontology is involved; but this is because the supply of sentences whose truth conditions are to be given is finite.

Not that an infinity of sentences necessarily demands ontology. Given the finite supply of sentences with unstructured predicates that we have been imagining, it is easy to go on to infinity, by adding one or more iterable devices for constructing sentences from sentences, like negation, conjunction, or alternation. If ontology was not required to give the truth conditions for the simplest sentences, these devices will not call for more.

In general, however, semantically relevant structure is apt to demand ontology. Consider, for example, the view that quotations are to be treated as semantic atoms, on a par with proper names in lacking significant structure. Tarski says of this way of viewing quotation that it 'seems to be the most natural one and completely in accordance with the customary way of using quotation marks'.[6] He gives a model argument to show that quotation marks cannot be

[6] A. Tarski, 'The Concept of Truth in Formalized Languages', 160. For more on quotation see Essay 6.

treated as an ordinary functional expression since a quotation does not name an entity that is a function of anything named by what the quotation marks enclose. About this Tarski is certainly right, but the moral of the lesson cannot be that quotations are like proper names—not, anyway, if a Tarski-style theory of truth can be given for a language containing quotation. For clearly there are infintely many quotations.

One idea for a possible solution can be extracted from Quine's remark that quotations may be replaced by spelling (much the same is said by Tarski). Spelling does have structure. It is a way of giving a semantically articulate description of an expression by the use of a finite number of expressions: the concatenation sign, with associated parentheses, and (proper) names of the letters. Following this line, we should think of a quotation like ' "cat" ' as having a form more clearly given by ' "c"⌢"a"⌢"t" ', or, better still, by '((see⌢eh)⌢tee'. This idea works, at least up to a point. But note the consequences. We no longer view the quotation ' "cat" ' as unstructured; rather we are treating it as an abbreviation of a sort for a complex description. Not, however, as an arbitrary abbreviation to be specified for the case at hand, but as a *style* of abbreviation that can be expanded mechanically into a description that shows structure more plainly. Indeed, talk of abbreviation is misleading; we may as well say this theory treats quotations as complex descriptions.

Another consequence is that in giving structure to quotations we have had to recognize in quotations repeatable and independent 'words': names of the individual letters, and the concatenation sign. These 'words' are, of course, finite in number—that was required—but they also reveal an ontological fact not apparent when quotations were viewed as unstructured names, a commitment to letters. We get a manageable theory when we explain molecules as made from atoms of a finite number of kinds; but we also get atoms.

A more stirring example of how postulating needed structure in language can bring ontology in its wake is provided by Frege's semantics for the oblique contexts created by sentences about propositional attitudes. In Frege's view, a sentence like 'Daniel believes that there is a lion in the den' is dominated by the two-place predicate 'believes' whose first place is filled by the singular term 'Daniel' and whose second place is filled by a singular term that names a proposition or 'sense'. Taking this line not only requires us to treat sentences as singular terms, but to find entities for them to name. And more is to come. For clearly an infinite number of sentences may occupy the spot after 'Daniel believes that . . .' So if we are to provide a truth definition, we must discover semantic structure in these singular terms: it must be shown how they can be treated as descriptions of propositions. To avoid the absurdities that would ensue if the singular terms in a sentence had their usual reference, Frege takes them as referring instead to intensional entities. Analogous changes must come over the semantic features of predicates, quantifiers, and sentential connectives. So far, a theory of truth of the sort we have been looking for can handle the situation, but only by treating each word of the language as ambiguous, having one interpretation in ordinary contexts and another after 'believes that' and similar verbs. What is to the eye one word must, from the vantage point of this theory, be treated as two. Frege appreciated this, and held the ambiguity against natural language; Church, in the artificial languages of 'A Formulation of the Logic of Sense and Denotation', eliminated the ambiguity by introducing distinct expressions, differing in subscript.[7]

Frege suggested that with each addition of a verb of propositional attitude before a referring expression that expression comes to refer to an entity of a higher semantical level. Thus every word and sentence is infinitely many-ways ambiguous; on Church's theory there will be an infinite basic vocabulary. In neither case is it possible to provide a theory of truth of the kind we want.

Frege was clear on the need, if we are to have a systematic theory, to view the truth value of each sentence as a function of the semantic roles of its parts or aspects, far clearer than anyone who went before, and clearer than most who followed. What Frege did not appreciate, as this last example brings out, was the additional restraints, in particular to a finite vocabulary, that flow from the demand for a comprehensive theory of truth. Frege brought semantics to a point where the demand was intelligible and even, perhaps, satisfiable; but it did not occur to him to formulate the demand.

Let us take a closer look at the bootstrap operation that enables us to bring latent structure to light by characterizing a truth predicate. Early steps may be illustrated by as simple a sentence as 'Jack and Jill went up the hill'—under what conditions is this sentence true? The challenge lies in the presence in the sentence of an

[7] A. Church, 'A Formulation of the Logic of Sense and Denotation'.

iterative device—conjunction. Clearly we can go on adding phrases like 'and Mary' after the word 'Jill' *ad libitum*. So any statement of truth conditions for this sentence must bear in mind the infinity of sentences, generated by the same device, that lie waiting for treatment. What is called for is a recursive clause in the truth theory that can be called into play as often as needed. The trick, as we all know, is to define truth for a basic, and finite, stock of simplest sentences, such as 'Jack went up the hill' and 'Jill went up the hill', and then make the truth conditions of 'Jack and Jill went up the hill' depend on the truth conditions, of the two simple sentences. So we get:

> 'Jack and Jill went up the hill' is true if and only if Jack went up the hill and Jill went up the hill.

as a consequence of a theory of truth. On the left, a sentence of the vernacular, its structure transparent or not, is described; on the right of the 'if and only if' a sentence of that same vernacular, but a part of the vernacular chosen for its ability to make explicit, through repeated applications of the same simple devices, the underlying semantic structure. If a theory of truth yields such a purified sentence for every sentence in the language, the portion of the total language used on the right may be considered a canonical notation. Indeed, with symbols substituted for some words, and grouping made plain by parentheses or some equivalent device, the part of the language used in stating truth conditions for all sentences may become indistinguishable from what is often called a formalized or artificial language. It would be a mistake, however, to suppose that it is essential to find such a canonical subdivision of the language. Since 'and' may be written between sentences in English, we take the easy route of transforming 'Jack and Jill went up the hill' into 'Jack went up the hill and Jill went up the hill' and then giving the truth conditions of the latter in accord with a rule that says a conjunction of sentences is true if and only if each conjunct is. But suppose 'and' never stood between sentences; its role as sentential connective would still be recognized by a rule saying that a sentence composed of a conjunctive subject ('Jack and Jill') and a predicate ('went up the hill') is true if and only if the sentence composed of the first conjoined subject and the predicate, and the sentence composed of the second conjoined subject and the predicate, are true. The rule required is less perspicuous, and needs to be supplemented with others, to do the work of the simple original rule. But the point remains: canonical notation is a convenience we can get along without if need be. It is good, but not necessary, to bring logical form to the surface.

Similarly, it would greatly ease the treatment of negation if we could plausibly transform all sentences containing negation into sentences, recognizably the same in truth value, in which the negating phrase always governs a sentence (as with, 'it is not the case that'). But if this were not possible, negation would still be a sentential connective if the truth condition of a sentence like 'Coal is not white' were given by adverting to the truth condition of 'Coal is white'. ('Coal is not white' is true if and only if 'Coal is white' is not true.)

The issue of ontology is forced into the open only where the theory finds quantificational structure, and that is where the theory best accounts for the pattern of truth dependencies by systematically relating expressions to objects. It is striking how firmly the demand for theory puts to rest one ancient source of aporia: the question how to demonstrate the asymmetry, if any, of subject and predicate. As long as our attention is focused on single, simple sentences, we may wonder why an explanation of truth should involve predicates in ontology any less than singular terms. The class of wise objects (or the property of wisdom) offers itself as what might correspond to the predicate 'wise' in 'Socrates is wise' in much the same way Socrates corresponds to 'Socrates'. As pointed out above, no finite number of such sentences requires a theory of truth to bring ontology into the picture. When we get to mixed quantification and predicates of any degree of complexity, however, the picture changes. With complex quantificational structure, the theory will match up expressions with objects. But there is no need, as long as the underlying logic is assumed to be first order, to introduce entities to correspond to predicates. Recognition of this fact will not, of course, settle the question whether there are such things as universals or classes. But it does demonstrate that there is a difference between singular term and predicate; for large stretches of language, anyway, variables, quantifiers, and singular terms must be construed as referential in function; not so for predicates.

It is not always evident what the quantificational structure of a sentence in natural language is; what appear to be singular terms sometimes melt into something less ontic in implication when their logical relations with other sentences are studied, while the requirements of theory may suggest that a sentence plays a role which can be explained only by treating it as having a quantificational structure not apparent on the surface. Here is a familiar illustration.

What is the ontology of a sentence like:

'Jack fell down before Jack broke his crown'?

Jack and his crown seem to be the only candidates for entities that must exist if this sentence is to be true. And if, in place of 'before', we had 'and', this answer might satisfy us for the reason already explored: namely, that we can state, in a way that will work for endless similar cases, the truth conditions of the whole sentence 'Jack fell down *and* Jack broke his crown' on the basis just of the truth of the component sentences, and we can hope to give the truth conditions for the components without more ontology than Jack and his crown. But 'Jack fell down before Jack broke his crown' does not yield to this treatment, because 'before' cannot be viewed as a truth-functional semantic connective: to see this, reflect that for the sentence to be true, both component sentences must be true, but this is not sufficient for its truth, since interchanging the components will make it false.

Frege showed us how to cope with the case: we can formulate the truth conditions for the sentence 'Jack fell down before Jack broke his crown' as follows: it is true if and only if there exists a time t and there exists a time t' such that Jack fell down at t, Jack broke his crown at t', and t is before t'. So apparently we are committed to the existence of times if we accept any such sentence as true. And thinking of the holistic character of a truth definition, the discovery of hidden intology in sentences containing 'before' must carry over to other sentences: thus, 'Jack fell down' is true if and only if there exists a time t such that Jack fell down at t.

Now for a more disturbing example. Consider first 'Jack's fall caused the breaking of his crown.' Here it is natural to take 'Jack's fall' and 'the breaking of his crown' as singular terms describing events, and 'caused' as a two-place, or relational, predicate. But then, what is the semantic relation between such general terms as 'fall' in 'Jack's fall' or 'the fall of Jack' and such verbs as 'fell' in 'Jack fell'? For that matter, how does 'Jack's fall caused the breaking of his crown' differ, in its truth conditions, from 'Jack fell, which caused it to be the case that Jack broke his crown', where the phrase 'which caused it to be the case that' is, on the face of it, a sentential connective?

The correct theory of 'caused', as I have argued at more length elsewhere, is parallel to Frege's theory for 'before'.[8] I suggest that 'Jack fell down, which caused a breaking of his crown' is true if and only if there exist events *e* and *f* such that *e* is a fall Jack took, *f* is a breaking his crown suffered, and *e* caused *f*. According to this proposal, the predicate 'is a fall', true of events, becomes primary, and contexts containing the verb are derived. Thus 'Jack fell' is true if and only if there is a fall such that Jack took it, 'Jack took a walk' is true if and only if there is a walk that he took, and so on. On this analysis, a noun phrase like 'Jack's fall' becomes a genuine description, and what it describes is the one fall that Jack took.

One consideration that may help reconcile us to an ontology of particular events is that we may then dispense with the abstract ontology of times we just now tentatively accepted, for events are as plausibly the relata of the before-relation as times. Another consideration is that by recognizing our commitment to an ontology of events we can see our way to a viable semantics of adverbs and adverbial modification. Without events, there is the problem of explaining the logical relations between sentences like 'Jones nicked his cheek while shaving with a razor in the bathroom on Saturday', and 'Jones nicked his cheek in the bathroom', and 'Jones nicked his cheek'. It seems that some iterative device is at work; yet what, from a semantic point of view, can the device be? The books on logic do not say: they analyse these sentences to require relations with varying numbers of places depending on the number of adverbial modifications, but this leads to the unacceptable conclusion that there is an infinite basic vocabulary, and it fails to explain the obvious inferences. By interpreting these sentences as being about events, we can solve the problems. Then we can say that 'Jones nicked his cheek in the bathroom on Saturday' is true if and only if there exists an event that is a nicking of his cheek by Jones, *and* that event took place in the bathroom, *and* it took place on Saturday. The iterative device is now obvious: it is the familiar collaboration of conjunction and quantification that enables us to deal with 'Someone fell down and broke his crown'.

This device works, but as we have seen, it takes an ontology to make it work: an ontology including people for 'Someone fell down

[8] See Essay 7 in *Essays on Actions and Events*.

and broke his crown', an ontology of events (in addition) for 'Jones nicked his cheek in the bathroom on Saturday'. It is mildly ironic that in recent philosophy it has become a popular manœuver to try to *avoid* ontological problems by treating certain phrases as adverbial. One such suggestion is that we can abjure sense-data if we render a sentence like 'The mountain appears blue to Smith' as 'The mountain appears bluely to Smith'. Another similar idea is that we can do without an ontology of intensional objects by thinking of sentences about propositional attitudes as essentially adverbial: 'Galileo said that the earth moves' would then come out, 'Galileo spoke in a-that-the-earth-moves-fashion'. There is little chance, I think, that such adverbial clauses can be given a systematic semantical analysis without ontological entanglements.

There is a further, rather different, way in which a theory of truth may have metaphysical repercussions. In adjusting to the presence of demonstratives, and of demonstrative elements like tense, in a natural language, a theory of truth must treat truth as an attribute of utterances that depends (perhaps among other things) on the sentence uttered, the speaker, and the time. Alternatively, it may be possible to treat truth as a relation between speakers, sentences, and times. Thus an utterance of 'I am five feet tall' is true if spoken at some times in the lives of most people, and true if spoken at any time during a considerable span in the lives of a few. 'Your slip is showing' may be true when uttered by a speaker at a time when he faces west, though it might not have been true if he had faced north; and 'Hilary climbed Everest' was for a long time false, and is now forever true. Sentences without demonstrative elements cannot do the work of sentences with demonstrative elements, but if we are to have a theory of truth, we must be able to state, without the use of demonstratives, a rule that explains under what conditions sentences with demonstratives are true. Such rules will give the truth condition of sentences like 'Hilary climbed Everest' only by quantifying over utterances, speakers, and times, or, perhaps, events.

If explicit appeal must be made to speakers and their circumstances in giving a theory of truth, then on the assumption that the general features of language reflect objective features of the world, we must conclude that an intelligible metaphysics will assign a central place to the idea of people (= speakers) with a location in public space and time.

It should be clear that 'the method of truth' in metaphysics does not eliminate recourse to more standard, often essentially non-linguistic, arguments or decisions. What it is possible to do in a theory of truth, for example, depends to a large extent on the logical resources the theory itself deploys, and the theory cannot decide this for us. Nor, as we have seen, does the method suggest what truths, beyond those it counts as logical, we must accept as a condition of mutual understanding. What a theory of truth does is describe the pattern truth must make among the sentences, without telling us where the pattern falls. So, for example, I argue that a very large number of our ordinary claims about the world cannot be true unless there are events. But a theory of truth, even if it took the form I propose, would not specify which events exist, nor even that any do. However, if I am right about the logical form of sentences concerning change, then unless there are events, there are no true sentences of very common kinds about change. And if there are no true sentences about change, there are no true sentences about objects that change. A metaphysician who is willing to suppose no sentences like 'Vesuvius erupted in March 1944' or 'Caesar crossed the Rubicon' are true will not be forced by a theory of truth to admit the existence of events or even, perhaps, of people or mountains. But if he accepts that many such sentences are true (whichever they may be), then it is obvious that he must accept the existence of people and volcanoes; and, if I am right, the existence of events like eruptions and crossings.

The merit of the method of truth is not that it settles such matters once and for all, or even that it settles them without further metaphysical reflection. But the method does serve to sharpen our sense of viable alternatives, and gives a comprehensive idea of the consequences of a decision. Metaphysics has generality as an aim; the method of truth expresses that demand by requiring a theory that touches all the bases. Thus the problems of metaphysics, while neither solved nor replaced, come to be seen as the problems of all good theory building. We want a theory that is simple and clear, with a logical apparatus that is understood and justified, and that accounts for the facts about how our language works. What those facts are may remain somewhat in dispute, as will certainly the wisdom of various trade-offs as between simplicity and clarity. These questions will be, I do not doubt, the old questions of metaphysics in new dress. But the new dress is in many ways an attractive one.

15 *Reality without Reference*

It is difficult to see how a theory of meaning can hope to succeed that does not elucidate, and give a central role to, the concept of reference. On the other hand, there are weighty reasons for supposing that reference cannot be explained or analysed in terms more primitive or behavioural. Let me describe the dilemma more fully, and then say how I think a theory of truth in Tarski's style can help resolve it.

'Theory of meaning' is not a technical term, but a gesture in the direction of a family of problems (a problem family). Central among the problems is the task of explaining language and communication by appeal to simpler, or at any rate different, concepts. It is natural to believe this is possible because linguistic phenomena are patently supervenient on non-linguistic phenomena. I propose to call a theory a theory of meaning for a natural language *L* if it is such that (a) knowledge of the theory suffices for understanding the utterances of speakers of *L* and (b) the theory can be given empirical application by appeal to evidence described without using linguistic concepts, or at least without using linguistic concepts specific to the sentences and words of *L*. The first condition indicates the nature of the question; the second requires that it not be begged.

By a theory of truth, I mean a theory that satisfies something like Tarski's Convention T: it is a theory that by recursively characterizing a truth predicate (say 'is true in *L*') entails, for each sentence *s* of *L*, a metalinguistic sentence got from the form '*s* is true in *L* if and only if *p*' when '*s*' is replaced by a canonical description of a sentence of *L* and '*p*' by a sentence of the metalanguage that gives the truth conditions of the described sentence. The theory must be relativized to a time and a speaker (at least) to handle indexical expressions.

Nevertheless I shall call such theories *absolute* to distinguish them from theories that (also) relativize truth to an interpretation, a model, a possible world, or a domain. In a theory of the sort I am describing, the truth predicate is not defined, but must be considered a primitive expression.

We may take reference to be a relation between proper names and what they name, complex singular terms and what they denote, predicates and the entities of which they are true. Demonstratives will not enter the discussion, but their references would, of course, have to be relativized to a speaker and a time (at least).

Now back to the dilemma. Here is why it seems that we can't do without the concept of reference. Whatever else it embraces, a theory of meaning must include an account of truth—a statement of the conditions under which an arbitrary sentence of the language is true. For well-known reasons, such a theory can't begin by explaining truth for a finite number of simple sentences and then assign truth to the rest on the basis of the simples. It is necessary, and in any case would be wanted in a revealing story, to analyse sentences into constituent elements—predicates, names, connectives, quantifiers, functors—and to show how the truth value of each sentence derives from features of the elements and the composition of the elements in the sentence. Truth then clearly depends on the semantic features of the elements; and where the elements are names or predicates, what features can be relevant but the reference? Explaining the truth conditions of a sentence like 'Socrates flies' must amount to saying it is true if and only if the object referred to by 'Socrates' is one of the objects referred to by the predicate 'flies'.

A theory of truth of the kind mentioned above does show how the truth conditions of each sentence are a function of the semantic features of the items in a basic finite vocabulary. But such a theory does not, it is often said, explain the semantic features of the basic vocabulary. In a theory of truth, we find those familiar recursive clauses that specify, for example, that a conjunction is true if and only if each conjunct is true, that a disjunction is true if and only if at least one disjunct is true, and so on. (In fact the theory must explain how connectives work in open as well as closed sentences, and so the recursion will be applied to the *satisfaction* relation rather than directly to truth.)

The theory must, we know, entail a T-sentence even for the simplest cases, for example:

(T) 'Socrates flies' is true iff Socrates flies.

If nothing is said about the constituents, how does the theory handle such cases? Well, one way might be this: The basic vocabulary must be finite. In particular, then, there can be only a finite number of simple predicates and a finite number of proper names (unstructured singular terms, not counting variables). So it is possible to list every sentence consisting of a proper name and a basic predicate. It follows that a theory can entail every sentence like (T) by having every such sentence as an axiom. Clearly this method avoids (so far) any appeal to the concept of reference—and fails to throw any light on it.

Predicates come in any degree of complexity, since they can be built up from connectives and variables; and constant singular terms can be complex. So the method we were just exploring will fail to work in general. In the case of predicates, Tarski's method, as we know, involves appeal to the concept of satisfaction, a relation between predicates and *n*-tuples of entities of which the predicates are true (actually, sequences of such). Satisfaction is obviously much like reference for predicates—in fact we might define the reference of a predicate as the class of those entities that satisfy it. The trouble is, an absolute theory of truth doesn't really illuminate the relation of satisfaction. When the theory comes to characterize satisfaction for the predicate '*x* flies', for example, it merely tells us that an entity satisfies '*x* flies' if and only if that entity flies. If we ask for a further explanation or analysis of the relation, we will be disappointed.

The fact that an absolute definition of truth fails to yield an analysis of the concept of reference can be seen from this, that if one imagines a new predicate added to the language—or a language just like the old except for containing a single further predicate—the account of truth and satisfaction already given don't suggest *how to go on to the new case*. (This remark doesn't apply to the recursive clauses: they say *in general* when a conjunction is true, no matter what the conjuncts are.)

The fact that satisfaction, which we have been thinking of as recursively characterized, can be given an explicit definition (by the Frege–Dedekind technique) should not lead us into thinking a general concept has been captured. For the definition (like the recursion that serves it) will explicitly limit the application of satisfaction to a fixed finite list of predicates (and compounds of them). So if a theory (or definition) of satisfaction applies to a given language and then a new predicate, say '*x* flies', is added, it will follow that '*x* flies' is not satisfied by an object that flies—or by anything else.

Analogous remarks go for constant singular terms. Indeed if there are complex singular terms, it will be necessary to characterize a relation like reference, using recursive clauses such as: 'the father of' concatenated with a name α refers to the father of what α refers to. But for the underlying proper names, there will again be simply a list. What it is for a proper name to refer to an object will not be analysed.

The point I have just been labouring, that there is a clear sense in which an absolute theory of truth does not throw light on the semantic features of the basic vocabulary of predicates and names, is familiar. This complaint has often been conflated with another, that a Tarski-style theory of truth gives no insight into the concept of truth. It will do no harm if we accept the idea that in a theory of truth, the expression 'is true' (or whatever takes its place) is understood independently. The reason Convention T is acceptable as a criterion of theories is that (1) T-sentences are clearly true (pre-analytically)—something we could recognize only if we already (partly) understood the predicate 'is true', and (2) the totality of T-sentences fixes the extension of the truth predicate uniquely. The interest of a theory of truth, viewed as an empirical theory of a natural language, is not that it tells us what truth is in general, but that it reveals how the truth of every sentence of a particular *L* depends on its structure and constituents.

We do not need to worry, then, over the fact that a theory of truth does not fully analyse the pre-analytic concept of truth. The point may be granted without impunging the interest of the theory. (I'll come back to this.) There remains the claim, which I also grant, that the theory does not explain or analyse the concept of reference. And that seems a grievous failure, since it undermines the pretensions of the theory to give a complete account of the truth of sentences.

This property of a theory of truth has been pointed out to me with increasing persistence by a number of critics. Gilbert Harman makes the point in order to question whether a theory of truth can, as I have claimed, do duty for a theory of meaning.[1] He says, in effect,

[1] G. Harman, 'Meaning and Semantics'.

that a theory of truth can only be considered as giving the meaning of the logical constants—it gives the logical form of sentences, and to that extent their meaning—but it cannot put flesh on the bones. Hartry Field develops the ideas I have touched on in the past few pages, and concludes that a Tarski-style theory of truth is only part of a complete theory.[2] We must in addition, he thinks, add a theory of reference for predicates and proper names. (I have drawn on his presentation.) Related criticisms have been made by Kathryn Pyne Parsons, Hilary Putnam and Paul Benacerraf.[3]

I have been saying why it seems that we can't live without the concept of reference; now let me say why I think we should be reluctant to live with it. I am concerned with what I take to be, historically at least, the central problem of philosophy of language, which is how to explain specifically linguistic concepts like truth (of sentences or utterances), meaning (linguistic), linguistic rule or convention, naming, referring, asserting, and so on—how to analyse some or all of these concepts in terms of concepts of another order. Everything about language can come to seem puzzling, and we would understand it better if we could reduce semantic concepts to others. Or if 'reduce' and 'analyse' are too strong (and I think they are), then let us say, as vaguely as possible, understand semantic concepts in the light of others.

To 'live with' the concept of reference means, in the present context, to take it as a concept to be given an independent analysis or interpretation in terms of non-linguistic concepts. The question whether reference is explicitly definable in terms of other semantic notions such as that of satisfaction, or recursively characterizable, or neither, is not the essential question—the essential question is whether it is *the*, or at least one, place where there is direct contact between linguistic theory and events, actions, or objects described in non-linguistic terms.

If we could give the desired analysis or reduction of the concept of reference, then all would, I suppose, be clear sailing. Having explained directly the semantic features of proper names and simple predicates, we could go on to explain the reference of complex singular terms and complex predicates, we could characterize satisfaction (as a derivative concept), and finally truth. This picture of

[2] H. Field, 'Tarski's Theory of Truth'.

[3] K. P. Parsons, 'Ambiguity and the Theory of Truth'; H. Putnam, 'The Meaning of "Meaning"'; P. Benacerraf, 'Mathematical Truth'.

how to do semantics is (aside from details) an old one and a natural one. It is often called the Building-Block theory. It has often been tried. And it is hopeless.

We have to go back to the early British empiricists for fairly clear examples of building-block theories. (Berkeley, Hume, Mill.) The ambitious attempts at behaviouristic analyses of meaning by Ogden and Richards and Charles Morris are not clear cases, for these authors tended to blur the distinction between words and sentences ('Fire!' 'Slab!' 'Block!') and much of what they said really applies intelligibly only to sentences as the basic atoms for analysis. Quine, in Chapter II of *Word and Object*, attempts a behaviouristic analysis, but although his most famous example ('Gavagai') is a single word, it is explicitly treated as a sentence. Grice, if I understand his project, wants to explain linguistic meaning ultimately by appeal to non-linguistic intentions— but again it is the meanings of *sentences*, not of words, that are to be analysed in terms of something else.

The historical picture, much simplified, shows that as the problems became clearer and the methods more sophisticated, behaviourists and others who would give a radical analysis of language and communication have given up the building-block approach in favour of an approach that makes the sentence the focus of empirical interpretation.

And surely this is what we should expect. Words have no function save as they play a role in sentences: their semantic features are abstracted from the semantic features of sentences, just as the semantic features of sentences are abstracted from *their* part in helping people achieve goals or realize intentions.

If the name 'Kilimanjaro' refers to Kilimanjaro, then no doubt there is *some* relation between English (or Swahili) speakers, the word, and the mountain. But it is inconceivable that one should be able to explain this relation without first explaining the role of the word in sentences; and if this is so, there is no chance of explaining reference directly in non-linguistic terms.

It is interesting that Quine in Chapter II of *Word and Object* makes no use of the concept of reference, nor does he try to construct it. Quine stresses the indeterminacy of translation, and of reference. He argues that the totality of evidence available to a hearer determines no unique way of translating one man's words into another's; that it does not even fix the *apparatus* of reference (singular terms, quantifiers, and identity). I think Quine understates his case. If it is true that the allowable evidence for interpreting a language is summed up when we know what the acceptable translation manuals from His to Ours are, then the evidence is irrelevant to questions of reference and of ontology. For a translation manual is only a method of going from sentences of one language to sentences of another, and we can infer from it nothing about the relations between words and objects. Of course we know, or think we know, what the words in our own language refer to, but this is information no translation manual contains. Translation is a purely syntactic notion. Questions of reference do not arise in syntax, much less get settled.

Here then, in brief, is the paradox of reference: There are two approaches to the theory of meaning, the building-block method, which starts with the simple and builds up, and the holistic method, which starts with the complex (sentences, at any rate) and abstracts out the parts. The first method would be fine if we could give a non-linguistic characterization of reference, but of this there seems no chance. The second begins at the point (sentences) where we can hope to connect language with behaviour described in non-linguistic terms. But it seems incapable of giving a complete account of the semantic features of the parts of sentences, and without such an account we are apparently unable to explain truth.

To return to the central dilemma: here is how I think it can be resolved. I propose to defend a version of the holistic approach, and urge that we must give up the concept of reference as basic to an empirical theory of language. I shall sketch why I think we can afford to do this.

The argument against giving up reference was that it was needed to complete an account of truth. I have granted that a Tarski-style theory of truth does not analyse or explain either the pre-analytic concept of truth or the pre-analytic concept of reference: at best it gives the extension of the concept of truth for one or another language with a fixed primitive vocabulary. But this does not show that a theory of absolute truth cannot explain the truth of individual sentences on the basis of their semantic structure; all it shows is that the semantic features of words cannot be made basic in interpreting the theory. What is needed in order to resolve the dilemma of reference is the distinction between explanation *within* the theory and explanation *of* the theory. Within the theory, the conditions of truth of a sentence are specified by adverting to postulated structure and semantic concepts like that of satisfaction or reference. But when it comes to interpreting the theory as a whole, it is the notion of truth, as applied to closed sentences, which must be connected with human ends and activities. The analogy with physics is obvious: we explain macroscopic phenomena by postulating an unobserved fine structure. But the theory is tested at the macroscopic level. Sometimes, to be sure, we are lucky enough to find additional, or more direct, evidence for the originally postulated structure; but this is not essential to the enterprise. I suggest that words, meanings of words, reference, and satisfaction are posits we need to implement a theory of truth. They serve this purpose without needing independent confirmation or empirical basis.

It should now be clear why I said, in the opening paragraph, that a theory of truth of the right sort could help resolve the apparent dilemma of reference. The help comes from the fact that a theory of truth helps us answer the underlying question how communication by language is possible: such a theory satisfies the two requirements we placed on an adequate answer (in the second paragraph). The two requirements relate directly to the distinction just made between explaining something in terms of the theory, and explaining why the theory holds (i.e., relating it to more basic facts).

Let us take the second condition first. How can a theory of absolute truth be given an empirical interpretation? What is essential in the present context is that the theory be related to behaviour and attitudes described in terms not specific to the language or sentence involved. A Tarski-style truth theory provides the obvious place to establish this relation: the T-sentences. If we knew all of these were true, then a theory that entailed them would satisfy the formal requirement of Convention T, and would give truth conditions for every sentence. In practice, we should imagine the theory builder assuming that some T-sentences are true on the evidence (whatever that is), building a likely theory, and testing further T-sentences to confirm, or supply grounds for modifying, the theory. A typical T-sentence, now relativized to time, might be:

> 'Socrates is flying' is true (in Smith's language) at *t* iff Socrates is flying at *t*.

Empirically, what we need is a relation between Smith and the sentence 'Socrates is flying' which we can describe in non-question-begging terms and which holds when and only when Socrates is flying. The theory will, of course, contain a recursion on a concept like satisfaction or reference. But these notions we must treat as theoretical constructs whose function is exhausted in stating the truth conditions for sentences. Similarly, for that matter, for the logical form attributed to sentences, and the whole machinery of terms, predicates, connectives, and quantifiers. None of this is open to direct confrontation with the evidence. It makes no sense, on this approach, to complain that a theory comes up with the right truth conditions time after time, but has the logical form (or deep structure) wrong. We should take the same view of reference. A theory of this kind does not, we agreed, explain reference, at least in this sense: it assigns no empirical content directly to relations between names or predicates and objects. These relations are given a content *indirectly* when the T-sentences are.

The theory gives up reference, then, as part of the cost of going empirical. It can't, however, be said to have given up ontology. For the theory relates each singular term to some object or other, and it tells what entities satisfy each predicate. Doing without reference is not at all to embrace a policy of doing without semantics or ontology.

I have not said what is to count as the evidence for the truth of a T-sentence.[4] The present enterprise is served by showing how the theory can be supported by relating T-sentences, and nothing else, to the evidence. What is clear is that the evidence, whatever it is, cannot be described in terms that relate it in advance to any particular language, and this suggests that the concept of truth to which we appeal has a generality that the theory cannot hope to explain.

Not that the concept of truth that is used in T-sentences can be explicitly defined in non-semantic terms, or reduced to more behaviouristic concepts. Reduction and definition are, as I said at the beginning, too much to expect. The relation between theory and evidence will surely be much looser.

A general and pre-analytic notion of truth is presupposed by the theory. It is because we have this notion that we can tell what counts as evidence for the truth of a T-sentence. But the same is not required of the concepts of satisfaction and reference. Their role is theoretical, and so we know all there is to know about them when we know how they operate to characterize truth. We don't need a

[4] See Essays 9–11.

general *concept* of reference in the construction of an adequate theory.

We don't need the concept of reference; neither do we need reference itself, whatever that may be. For if there is one way of assigning entities to expressions (a way of characterizing 'satisfaction') that yields acceptable results with respect to the truth conditions of sentences, there will be endless other ways that do as well. There is no reason, then, to call any one of these semantical relations 'reference' or 'satisfaction'.[5]

How can a theory of absolute truth give an account of communication, or be considered a theory of meaning? It doesn't provide us with the materials for defining or analysing such phrases as 'means', 'means the same as', 'is a translation of', etc. It is wrong to think that we can *automatically* construe T-sentences as 'giving the meaning' of sentences if we put no more constraint on them than that they come out true.

The question to ask is whether someone who knows a theory of truth for a language L would have enough information to interpret what a speaker of L says. I think the right way to investigate this question is to ask in turn whether the empirical and formal constraints on a theory of truth sufficiently limit the range of acceptable theories. Suppose, for example, that *every* theory that satisfied the requirements gave the truth conditions of 'Socrates flies' as suggested above. Then clearly to know the theory (and to know *that* it is a theory that satisfies the constraints) is to know that the T-sentence *uniquely* gives the truth conditions of 'Socrates flies'. And this *is* to know enough about its role in the language.

I don't for a moment imagine such uniqueness would emerge. But I do think that reasonable empirical constraints on the interpretation of T-sentences (the conditions under which we find them true), plus the formal constraints, will leave enough invariant as between theories to allow us to say that a theory of truth captures the essential role of each sentence. A rough comparison may help give the idea. A theory of measurement for temperature leads to the assignment to objects of numbers that measure their temperature. Such theories put formal constraints on the assignments, and also must be tied empirically to qualitatively observable phenomena. The numbers assigned are not uniquely determined by the constraints.

[5] See J. Wallace, 'Only in the Context of a Sentence do Words Have Any Meaning'.

But the *pattern* of assignments is significant. (Fahrenheit and Centigrade temperature are linear transformations of each other; the assignment of numbers is unique up to a linear transformation.) In much the same way, I suggest that what is invariant as between different acceptable theories of truth is meaning. The meaning (interpretation) of a sentence is given by assigning the sentence a semantic location in the pattern of sentences that comprise the language. Different theories of truth may assign different truth conditions to the same sentence (this is the semantic analogue of Quine's indeterminacy of translation), while the theories are (nearly enough) in agreement on the roles of the sentences in the language.

The central idea is simple. The building-block theory, and theories that try to give a rich content to each sentence directly on the basis of non-semantic evidence (for example the intentions with which the sentence is typically uttered), try to move too far too fast. The present thought is rather to expect to find a minimum of information about the correctness of the theory at each single point; it is the potential infinity of points that makes the difference. A strong theory weakly supported, but at enough points, may yield all the information we need about the atoms and molecules—in this case, the words and sentences.

In a nutshell: we compensate for the paucity of evidence concerning the meanings of individual sentences not by trying to produce evidence for the meanings of words but by considering the evidence for a theory of the language to which the sentence belongs. Words and one or another way of connecting them with objects are constructs we need to implement the theory.

This conception of how to do theory of meaning is essentially Quine's. What I have added to Quine's basic insight is the suggestion that the theory should take the form of a theory of absolute truth. If it does take this form, we can recover a structure of sentences as made up of singular terms, predicates, connectives, and quantifiers, with ontological implications of the usual sort. Reference, however, drops out. It plays no essential role in explaining the relation between language and reality.

16 *The Inscrutability of Reference*

Quine's thesis of the unscrutability of reference is that there is no way to tell what the singular terms of a language refer to, or what its predicates are true of, at least no way to tell from the totality of behavioural evidence, actual and potential, and such evidence is all that matters to questions of meaning and communication. Inscrutability of reference follows from considerations made plain in *Word and Object*, but the term and the thesis became central only in 'Ontological Relativity'. The thesis is important because the indeterminacy of translation follows directly from it, and it is easier to give clear grounds for the thesis than for some other forms of indeterminacy.

The clear grounds are provided by examples: Quine and others have shown how to construct systematic examples of alternative schemes of reference such that, if one of them is in accord with all possible relevant evidence, others are. Quine argues that this fact should lead us to recognize that the relation of reference between objects and words (or their utterances) is relative to an arbitrary choice of a scheme of reference (or translation manual), and in fact relative to a further basic parameter. The argument for relativism puzzles me, for I find it impossible to formulate the relativized concept of reference in an acceptable way. In this paper I discuss my difficulties and offer Quine a way out of one of them. Of course, I hope that my way out is what Quine had in mind all along. In that case the exercise will be of no use to him, but will have been to me.

To make my general position clear from the start, I accept Quine's thesis of the unscrutability of reference and therefore of the indeterminacy of translation. And I think that I accept both these mainly on the basis of arguments that I have learned from Quine.

But I do not see how these arguments show reference to be relative in the way that Quine believes it is; indeed, I think Quine's own views undermine the idea that ontology can be relativized.

Where Quine mostly speaks of ontological relativity I have been using the phrase relativity of reference. My reason for the shift is that I want to concentrate on a particularly clear and simple kind of example of the inscrutability of reference. In this kind of example the total ontology is assumed to be fixed, but the truth of sentences is explained by matching up objects with words in different ways. Before I say more about such examples, let me emphasize the narrow scope of this paper by classifying the various kinds of indeterminacy that Quine discusses, for any of them can make reference inscrutable.

First, truth itself may be indeterminate; there may be a translation manual (or, better, for present purposes, a theory of truth) for a language that satisfies all relevant empirical constraints and makes a certain sentence true and another equally acceptable theory that does not make that sentence true (of course, there will have to be other differences too). I shall briefly discuss such cases later, but only by the way.

Second, logical form may be indeterminate: two satisfactory theories may differ in what they count as singular terms or quantifiers or predicates, or even with respect to the underlying logic itself.

Third, even if logical form and truth are fixed, acceptable theories may differ with respect to the references they assign to the same words and phrases. And here we can subdivide into cases where the total ontology differs and cases where it does not. I shall be almost exclusively concerned with the second subdivision of the third kind of indeterminacy.

I believe that all these sorts of indeterminacy are possible, but, except for the last one, the one I shall be discussing, I probably think the range of indeterminacy is less than Quine thinks it is. The first sort of indeterminacy I would reduce by a more far-reaching application of the principle of charity than Quine deems essential. The second sort of indeterminacy is automatically put under greater control if one insists, as I do, on a Tarski-style theory of truth as the basis of an acceptable translation manual. And variance in total ontology can, in my opinion, be made intelligible only to a limited degree; there is a principle of reverse charity that judges a theory better the more of its own resources it reads into the language for which it is a theory. I shall give no reasons for these views here, but neither shall I depend on them.

It is hard, however, to talk about reference without placing it in the context of a theory of truth. For a satisfactory account of reference is one that assigns an extension to each of the predicates and singular terms of a language, and this requires a recursive characterization that can handle quantifiers and perhaps functions. In any case, if the language has the resources we are certain to credit a natural language with, we will want to introduce a concept like that of satisfaction which is more general than reference and on the basis of which truth and reference (at least for predicates) can be defined. I shall assume in what follows that a relation of reference holds between names and what they might be said to name, singular terms and what they might be said to denote, and n-adic predicates and the n-tuples of which they might be said to be true, all these kinds of reference being knit together by a concept like that of satisfaction which yields a definition of truth for closed sentences. All these 'might be said to' phrases are, of course, in deference to the inscrutability of reference.

The simplest, least questionable way of showing that reference is inscrutable depends on the idea of a permutation of the universe, some one-to-one mapping of every object on to another. Let us suppose ϕ is such a permutation. If we have a satisfactory scheme of reference for a language that speaks of this universe, we can produce another scheme of reference by using the permutation: whenever, on the first scheme, a name refers to an object x, on the second scheme it refers to $\phi(x)$; whenever, on the first scheme, a predicate refers to (is true of) each thing x such that Fx, on the second scheme it refers to each thing x such that $F\phi(x)$. Assuming that reference is geared to an appropriate characterization of a relation like satisfaction in each case, it is easy to see that the truth conditions the second scheme assigns to a sentence will in every case be equivalent to the truth conditions assigned to that sentence by the first scheme.[1] Indeed, as Wallace points out, we can even have our two theories of truth and reference

[1] For further details, and answers to various difficulties, see J. Wallace, 'Only in the Context of a Sentence do Words Have Any Meaning'. See also Hartry Field's two articles, 'Quine and the Correspondence Theory' and 'Conventionalism and Instrumentalism in Semantics'. The present paper has been much influenced by the articles of Wallace and Field.

yield identical truth conditions for all sentences if we add to the theory the assumptions that led us to believe ϕ was a permutation of the universe.

The earliest examples known to me of the use of permutations of the universe to illustrate the slippery nature of reference are in a review by Richard Jeffrey.[2] His main target is not reference, however, and his examples do not exactly fill our bill. Further examples can be found in the works referred to above by John Wallace and Hartry Field. Here is a simple illustration: suppose every object has one and only one shadow.[3] Then we may take ϕ to be expressed by the words 'the shadow of'. On a first theory, we take the name 'Wilt' to refer to Wilt and the predicate 'is tall' to refer to tall things; on the second theory, we take 'Wilt' to refer to the shadow of Wilt and 'is tall' to refer to the shadows of tall things. The first theory tells us that the sentence 'Wilt is tall' is true if and only if Wilt is tall; the second theory tells that 'Wilt is tall' is true if and only if the shadow of Wilt is the shadow of a tall thing. The truth conditions are clearly equivalent. If one does not mind speaking of facts, one might say that the same fact makes the sentence true in both cases.

I assume that there are permutations of the requisite kind that demand no fiction. Another assumption that is clearly needed if we are to conclude to the inscrutability of reference is that if some theory of truth (or translation or interpretation) is satisfactory in the light of all relevant evidence (actual or potential) then any theory that is generated from the first theory by a permutation will also be satisfactory in the light of all relevant evidence. Of course many philosophers reject this assumption, but since it is a matter on which I agree with Quine, I shall not argue for it here. The crucial point on which I am with Quine might be put: all the evidence for or against a theory of truth (interpretation, translation) comes in the form of facts about what events or situations in the world cause, or would cause, speakers to assent to, or dissent from, each sentence in the speaker's repertoire. We probably differ on some details. Quine describes the events or situations in terms of patterns of stimulation, while I prefer a description in terms more like those of the sentence being studied; Quine would give more weight to a grading of sentences in terms of observationality than I would; and where he

[2] R. Jeffrey, review of *Logic, Methodology, and the Philosophy of Science*.

[3] For the example to work properly, everything must be, as well as have, a shadow.

likes assent and dissent because they suggest a behaviouristic test, I despair of behaviourism and accept frankly intensional attitudes toward sentences, such as holding true. So far as I can see, none of these differences matters to the argument for the inscrutability of reference. What matters is that what causes the response or attitude of the speaker is an objective situation or event, and that the response or attitude is directed to a sentence or the utterance of a sentence. As long as we hold to this, there can be no relevant evidence on the basis of which to choose between theories and their permutations.

Does the inscrutability of reference, understood and defended as just now, support the idea that reference should be relativized? It certainly suggests it. For we cannot rest with a conclusion that allows us to accept both that 'Wilt' refers to Wilt and that 'Wilt' refers to the shadow of Wilt. We can without contradiction accept both only if both can be true, and clearly this is not the case.

It is possible to solve the problem without relativizing. All we need is to make clear that 'refers' is being used in two ways. Subscripts would put things right. 'Wilt' refers_1 to Wilt, and 'Wilt' refers_2 to the shadow of Wilt; this is a conjunction that can be true, and is under circumstances we have sketched. So far, no relativization, though the use of the same word with different subscripts hints at a common feature that relativization might make explicit. A further reason for wanting to relativize reference is that we want to say something like this: relative to our first way of doing things, the right answer to the question, what 'Wilt' refers to is Wilt; relative to the second way, it refers to the shadow of Wilt. Since ways of doing things are not attractive entities to quantify over, Quine proposes that reference be relativized to translation manuals. Hartry Field has pointed out that this won't work.[4] For the natural way to state the conditions under which 'x refers to y relative to TM' holds is this: TM translates x as 'y'. This suggestion must be rejected because you cannot quantify into quotation marks.

I think there is a general reason why reference cannot be relativized in the way that Quine wants, so it is fruitless to try to improve on the formulation we just rejected. When I say 'in the way that Quine wants', what I am really objecting to is not *any* way of relativizing reference, for I shall in the end propose a way. What I

[4] H. Field, 'Quine and the Correspondence Theory', 206.

object to is the idea that reference can be relativized in such a way as to fix ontology. It is ontological relativity that I do not understand. Suppose we could fix the ontology of 'refers' by relativizing it. Then we would have fixed the ontology of the language or speaker we were using the word 'refers' to characterize. It may be said: but the fixing is only relative to an arbitrary choice. That choice is not dictated by any relevant evidence. Hence the inscrutability. This reply misses the point of the difficulty. The fixing of reference and ontology for the object language has been done on the basis of an arbitrary choice; but the arbitrary choice succeeds in doing this only if the relativized 'refers' of the metalanguage has somehow been nailed down. And this is what we argued cannot be done for any language.

It is perhaps this line of reasoning that makes Quine say that reference and ontology are *doubly* relative, once to a choice of a manual of translation and once to some background theory or language.[5] Since relativizing reference in the metalanguage cannot pin down reference and ontology for the object language unless the relativized reference predicate of the metalanguage has an unambiguous semantics, Quine sees an infinite hierarchy of theories or languages each of which tries (vainly) to stabilize the reference scheme of the language for which it provides the theory. Quine compares the relativity of reference to a background theory to the relativity of location to a co-ordinate system. But the comparison fails. In the case of location the case is clear. It never makes sense simply to ask where an object is, but it does make sense to ask where an object is relative to other objects—a co-ordinate system (of course the relativity may go unmentioned because some frame of reference is assumed). The relativized question ('Where is Bronk's house in the address system of the Bronx?') is clear and answerable, and the answer is complete. It has no further hidden parameter. We can, of course, go on to ask another, similar question: 'Where is the Bronx?' And this question in turn makes no sense until relativized. But once relativized, it is clear and answerable. No predicate with an extra place is hidden behind the relativized location predicate.

How then is it possible to express the relativism of reference to a background language? Quine allows that 'we can and do talk meaningfully and distinctively of rabbits and parts, numbers and

[5] Quine sometimes says 'background theory', sometimes 'language'. See 'Ontological Relativity', 48, 54, 55, 67.

formulas', but only relative to our own language.[6] This perhaps suggests that the relativism can be put into our own language. It is easy to see, however, why this is impossible: if the question of reference in my own language is unsettled, it cannot help to try in that same language to say something to resolve the question. If you understand me, you will rightly interpret my utterance of 'Wilt is tall'. If there is some trouble, I cannot help you by adding 'in English', since the same trouble would force me to add the same phrase to each successively enlarged version of my sentence. So if reference is relative to my frame of reference as already embedded in my own language, all that can be provided to give my words a reference is provided simply by my speaking my own language. The same must be true of my word 'reference' as applied to another language. But this is, again, just what Quine denies.

Quine compares the relativity of reference to a background language with the situation with respect to truth and satisfaction; there also, he reminds us, we have the prospect of a regress. We can define truth for L in M, but not in L, truth for M in M', but not in M, and so on. The same goes for reference.

The analogy seems to be faulty. What we cannot define in L is not made definable by adding a parameter. Nor is truth in L as defined in M somehow relative to truth in M as defined in M'. Truth is relative to an object language, but not to a metalanguage. The predicate 'is true in L' as it occurs in M does have a sense we can, if we want, specify in still another language. But how does this make truth in L relative to this third language—or even to M?

Some concept of conceptual relativism seems to inspire Quine's claim that reference, truth, and ontology must be relativized to a background theory or language. When he says that we can and do talk meaningfully and distinctively of rabbits and parts, but only relative to our frame of reference, the explanation of the claim does no more, in effect, than say that when we speak thus we must speak a language we know. This goes without saying, however, and neither invites nor allows an explicit statement by us of what our remarks are relative to. We can, as Quine insists, state the relativity if we back off to another language, but if this tactic is called for once, it is called for every time: an infinite regress. If this is the situation, ontology is not merely 'ultimately inscrutable',[7] any claim about

[6] Ibid., 48.

[7] Ibid., 51.

reference, however, many times relativized, will be as meaningless as 'Socrates is taller than'.

Quine is aware, of course, of the paradox in cultural relativism: elsewhere he writes that someone 'cannot proclaim cultural relativism without rising above it, and he cannot rise above it without giving it up'.[8] I would say the same about ontological relativism and the relativism of reference to a background theory or language. Quine once again leaves us hanging, on the issue of relativism, in the essay from which I have just quoted. He goes to lengths to convince us that there may be two theories such that they imply all and only true observation sentences, are equally simple, and yet logically incompatible.[9] Thus truth is apparently relative to a theory. But Quine settles, at the end, for a 'frank dualism' in which distinctive signs are used to state the theories. The theories turn out to be irreducible one to the other, but unconflicting. No relativism remains.

The lesson, in my opinion, is that we cannot make sense of truth, reference, or ontology relativized to a background theory or language. The trouble is not that we must start a regress we cannot finish. The trouble is that we do not understand the first step. Let me put my general argument for this once more, and in a slightly different way. Suppose that *B* states, in his own words, two theories of truth for a speaker *A*. One of his theories entails, or specifies, that 'Wilt' refers to Wilt, the other that 'Wilt' refers to the shadow of Wilt. Now *C* comes along and tries to work out a theory to interpret what *B* says. He discovers very soon, naturally, that the word 'refers' must be given two interpretations: *C* mutters to himself that *B* should have used different words, or somehow have relativized the predicate. In any case, *C* must, if he is to understand *B*, give different truth conditions to *B*'s two sentences about *A*'s word 'Wilt', whether or not *B* bothers to use two words, or to make explicit mention of a further parameter. But can *C* non-arbitrarily give the extension of *B*'s predicates, whether or not they are subscripted or relativized? No, for any satisfactory theory he has for understanding *B* can be transformed into endless equally satisfactory theories. *Nothing B* says can change this. *A* can talk distinctively and meaningfully about Wilt and shadows. *B* can talk distinctively and meaningfully about two different relations between *A*'s words and objects. But at no

[8] W. V. Quine, 'On Empirically Equivalent Systems of the World', 328.
[9] W. V. Quine, 'Ontological Relativity', 51.

point has anyone been able uniquely to specify the objects of which a predicate is true, no matter how arbitrarily or relatively.

Perhaps someone (not Quine) will be tempted to say, 'But at least the speaker knows what he is referring to.' One should stand firm against this thought. The semantic features of language are public features. What no one can, in the nature of the case, figure out from the totality of the relevant evidence cannot be part of meaning. And since every speaker must, in some dim sense at least, know this, he cannot even intend to use his words with a unique reference, for he knows that there is no way for his words to convey this reference to another.

To recapitulate: the argument for the inscrutability of reference has two steps. In the first step we recognize the empirical equivalence of alternative reference schemes. In the second step we show that, although an interpreter of the schemer can distinguish between the schemer's schemes, the existence of equivalent alternative schemes for interpreting the schemer prevents the interpreter from uniquely identifying the reference of the schemer's predicates, in particular his predicate 'refers' (whether or not indexed or relativized). What an interpreter cannot, on empirical grounds, decide about the reference of the schemer's words cannot be an empirical feature of those words. So those words do not, even when chosen from among arbitrary alternatives, uniquely determine a reference scheme. Hence the inscrutability of reference. Ontological relativity does not follow, since it suggests that, when enough decisions, arbitrary or otherwise, have been made, unique reference is possible, contrary to our argument for the inscrutability of reference.

If we could, by some choice, arbitrary or not, uniquely fix reference, the field would be open to Field (and others) to declare that the choice was not arbitrary and hence reference not inscrutable. And this Field has in fact declared. According to Field, the role of reference is not exhausted by its contribution to the truth conditions of sentences.[10] As a consequence there are grounds for making a choice among the theories of reference that yield equivalent truth conditions for sentences. Among those grounds are certain causal connections between names and what they refer to and (if Putnam is right) between predicates and what they are true of.

[10] See the two articles by H. Field mentioned in footnote 1 above.

The topic is a large and intricate one, and I can say only a little about it here. It is not, moreover, a matter that divides my views from Quine's. Nevertheless, it is certainly germane to the present discussion. Field has, in my opinion, confused the issue by assuming that if there are, say, causal connections between words and objects then it cannot be the case that theories of truth (and meaning and reference) are tested solely by evidence concerning sentences and their utterances. He is entranced by the fact that standard theories of truth explain what makes a sentence true by assigning semantic roles to the parts. So, he believes, we must give an independent account of the semantic properties of the parts (of reference, naming, satisfaction) if a theory of truth is to be true to its explanatory claims and virtues. This way of putting things fails to note the difference between explaining truth, given the theory, and providing evidence that the theory is true of some speaker or community. It is perfectly consistent to hold that a theory is testable only at the level of sentences while explaining the features of sentences on the basis of an inner structure. But if so then it is the semantic features of sentences (for example truth) that should be viewed as most directly connected with the evidence, while the semantic features of words, however posited, would do their work provided they explained the features of sentences: to take this stance is to assume that truth is easier to connect with non-linguistic evidence than reference. For this I think we can make a good case.[11]

But not now. My theme now is that, even if words do have, say, causal connections with what they refer to, this does not mean that the adequacy of a theory of truth is not to be tested at the sentential level. Suppose some causal theory of names is true. How are we to establish this fact as holding for the language of a particular speaker or community? Only, I suggest, by finding that the causal theory accounts for the linguistic behaviour, potential and actual, of the speakers. This behaviour is primarily concerned with sentences and their utterances. For imagine there was evidence that words, when *not* employed in sentences, had a certain causal connection with objects but that this fact had no bearing on how the words were used in sentences. Surely we would conclude that the first phenomena were irrelevant to an account of the language. If this is right,

[11] I try to make the case in Essay 15. Also see J. Wallace, 'Only in the Context of a Sentence do Words Have Any Meaning'.

determining that a causal theory of reference is true for a speaker must depend on evidence drawn from how sentences are viewed or used as much as any other theory of language.

Let us assume that there are causal connections between words, or their uses, and objects. Then a sentence like 'Wilt is tall' may be true only if, among other things, an utterance of the word 'Wilt' in this verbal context is causally connected with Wilt. Have we now, at least partially, saved the schemer from a merely arbitrary choice of schemes of reference for the speaker?

It seems not, for reasons explained by Wallace and Field. For suppose, as before, that ϕ is a permutation of the universe, and that Cx,y is an appropriate causal relation between a word and object. One good theory says that 'Wilt' refers to Wilt only if C 'Wilt', Wilt (I'm fudging on the difference between words and utterances of them), while another empirically indistinguishable theory says that 'Wilt' refers to ϕ (Wilt) only if C 'Wilt', ϕ (Wilt). Of course, neither 'refers' nor 'C' can have the same interpretation in both theories, but their subscripted counterparts are easy to specify. The two theories are clearly distinct, for 'refers$_1$' cannot have the same extension as 'refers$_2$' and 'C_1' cannot have the same extension as 'C_2'. But given that the first theory is satisfactory, we can define 'refers$_2$' and 'C_2' on the basis of 'refers$_1$' and 'C_1' and 'ϕ' to make the second theory empirically equivalent.

No causal theory, nor other 'physicalistic' analysis of reference, will affect our argument for the inscrutability of reference, at least as long as we allow that a satisfactory theory is one that yields an acceptable explanation of verbal behaviour and dispositions. For the constraints on the relations between reference and causality (or whatever) can always be equivalently captured by alternative ways of matching up words and objects. The interpreter of the schemer will, as before, be able to tell that the schemer's schemes are different from one another, but he will not be able to pick out a unique correct way of matching the schemer's words and objects. It follows that the schemer cannot have used words that determined a unique scheme. Reference remains inscrutable.

Causal theories of naming or of reference more generally are provocative, and in some version may well be true. If so, earlier concepts of naming and reference must be revised. But the question whether a causal theory of reference is true is independent, if I am right, of two issues with which it is often thought to be connected:

the issue of whether or not a theory of truth is to be tested by the reactions or attitudes of speakers to sentences and the issue of the inscrutability of reference.

What we have shown, or tried to show, is not that reference is not relative but that there is no intelligible way of relativizing it that justifies the concept of ontological relativity. The relativization must appear in the language in which the relativized predicate occurs (and hence cannot be to that language or to a theory for that language), and we cannot claim that it settles the question of reference in any language. But there *is* something to settle, and relativization is the only attractive way to settle it. On the one hand, all 'schemes of reference' that are acceptable for a speaker or community have important elements in common: they lead to equivalent truth conditions for all sentences, and perhaps they connect reference with causal or other matters in a specifiable pattern. On the other hand, such schemes are distinguishable one from another, not, indeed, on the basis of evidence but because they utilize predicates that cannot have the same extension. These two facts taken together strongly suggest a relativized concept. Besides, there is the question to which we can give an answer: 'What scheme are you using in giving that interpretation of the speaker's words?' If I interpret (translate) a speaker's word 'Wilt' as referring to the shadow of Wilt, I will probably need to explain how I am interpreting his predicate 'is tall' in accord with a scheme congenial to my interpretation of 'Wilt'. In some sense or other, my interpretation or translation is relative to, or based on, a specific scheme. The scheme may not settle matters of reference, but it does settle how I answer all sorts of questions about what the speaker means or refers to by a word or sentence. Quine frequently puts the matter just this way: 'It makes no sense to say what the objects of a theory are, beyond saying how to interpret or reinterpret that theory in another,'[12] or, again, 'What makes sense is to say not what the objects of a theory are, absolutely speaking, but how one theory of objects is interpretable or reinterpretable in another.' The second quotation bothers me, since the 'absolutely speaking' suggests that there is a way of relatively speaking that will decide, perhaps arbitrarily, what the *objects* are, and this I have strongly denied. But we cannot deny that, given a scheme of interpretation or translation, we have decided what *words* we can use

[12] W. V. Quine, 'Ontological Relativity', 50.

in our own language to interpret the words of a speaker. Is there an unobjectionable way to mark the relativity of our interpretation to our scheme? I think there is.

All that we can say gets fixed by the relativization is the way we answer questions about reference, not reference itself. So it seems to me the natural way to explain the sometimes needed explicit relativization is a familiar one: we take the speaker to be speaking one language or another. If we take his word 'rabbit' to refer to rabbits, we take him to be speaking one language. If we take his word 'rabbit' to refer to things that are ϕ of rabbits, we take him to be speaking another language. If we decide to change the reference scheme, we decide that he is speaking a different language. In some cases the decisions is ours; some languages are identical in that their speakers' dispositions to utter sentences under specified conditions are identical. There is no way to tell which of these languages a person is speaking.

The point will seem less trivial when we reflect that an empirical theory of a person's language does not stand alone: it is part of a more general theory that includes a theory of his beliefs, desires, intentions, and perhaps more. If we change our interpretation of a person's words, then, given the same total evidence, we must also change the beliefs and desires we attribute to him. It is not strange that we can take the same person to be speaking different languages, provided we can make compensatory adjustments in the other attitudes we attribute to him.

The issue will be clearer if we temporarily abandon the restriction we imposed at the onset to consider only theories that left the truth of sentences unaltered. There are often cases, I believe with Quine, when the totality of relevant evidence in a person's behaviour is equally well handled by each of two theories of truth, provided we make compensating adjustments in our theory of his beliefs and other attitudes, and yet where on one theory a particular sentence is interpreted in such a way as to make it true, and on the other not. Ian Hacking once put this puzzle to me: how can two theories of truth both be acceptable if one theory makes a certain utterance true and the other does not? Isn't this a contradiction? It is not a contradiction if the theories are relativized to a language, as all theories of truth are. Our mistake was to suppose there is a unique language to which a given utterance belongs. But we can without paradox take that utterance to belong to one or another language, provided we make allowance for a shift in other parts of our total theory of a person.

There is, then, a reasonable way to relativize truth and reference: sentences are true, and words refer, relative to a language. This may appear to be a familiar and obvious point, and in a way it certainly is. But there are some subtleties in how we understand it. It is not, for example, an empirical claim that 'Wilt' refers to Wilt in *L*. For if this were empirical, *L* would have to be characterized as *the* language spoken by some person or persons at a given time. Such a characterization would not serve our purpose, since we admit that it is not entirely an empirical question what language a person speaks; the evidence allows us some choice in languages, even to the point of allowing us to assign conflicting truth conditions to the same sentence. But even if we consider truth invariant, we can suit the evidence by various ways of matching words and objects. The best way of announcing the way we have chosen is by naming the language; but then we must characterize the language as one for which reference, satisfaction, and truth have been assigned specific roles. An empirical question remains, to be sure: is this language one that the evidence allows us to attribute to this speaker?

What permits us to choose among various languages for a speaker is the fact that the evidence—attitudes or actions directed to sentences or utterances—bears not only on the interpretation of speech but also on the attribution of belief, wants, and intentions (and no doubt other attitudes).[13] The evidence allows us a choice among languages because we can balance any given choice by an appropriate choice of beliefs and other attitudes. This suggests one more way we could relativize a theory of truth or reference: given certain assumptions about the nature of belief and other attitudes, we could show that, once we have decided what a person's attitudes are, the choice of a language is no longer up for grabs. Given a comprehensive account of belief, desire, intention, and the like, it is an empirical question what language a person speaks. And so we have, at last, a rather surprising way of making significant sense of the question what a word refers to.

It would, I hope it is clear, be a mistake to suppose that we somehow could first determine what a person believes, wants, hopes for, intends, and fears and then go on to a definite answer to the question

[13] This is a point emphasized in Essays 10 and 11.

what his words refer to. For the evidence on which all these matters depend gives us no way of separating out the contributions of thought, action, desire, and meaning one by one. Total theories are what we must construct, and many theories will do equally well. This is to state once more the thesis of the inscrutability of reference, but it is also to hint at the reason for it.

LIMITS OF THE LITERAL

17 *What Metaphors Mean*

Metaphor is the dreamwork of language and, like all dreamwork, its interpretation reflects as much on the interpreter as on the originator. The interpretation of dreams requires collaboration between a dreamer and a waker, even if they be the same person; and the act of interpretation is itself a work of the imagination. So too understanding a metaphor is as much a creative endeavour as making a metaphor, and as little guided by rules.

These remarks do not, except in matters of degree, distinguish metaphor from more routine linguistic transactions: all communication by speech assumes the interplay of inventive construction and inventive construal. What metaphor adds to the ordinary is an achievement that uses no semantic resources beyond the resources on which the ordinary depends. There are no instructions for devising metaphors; there is no manual for determining what a metaphor 'means' or 'says'; there is no test for metaphor that does not call for taste.[1] A metaphor implies a kind and degree of artistic success; there are no unsuccessful metaphors, just as there are no unfunny jokes. There are tasteless metaphors, but these are turns that nevertheless have brought something off, even if it were not worth bringing off or could have been brought off better.

This paper is concerned with that metaphors mean, and its thesis is that metaphors mean what the words, in their most literal interpretation, mean, and nothing more. Since this thesis flies in the face of contemporary views with which I am familiar, much of what I have to say is critical. But I think the picture of metaphor that

[1] I think Max Black is wrong when he says, 'The rules of our language determine that some expressions must count as metaphors.' ('Metaphor', 29.) There are no such rules.

emerges when error and confusion are cleared away makes metaphor a more, not a less, interesting phenomenon.

The central mistake against which I shall be inveighing is the idea that a metaphor has, in addition to its literal sense or meaning, another sense or meaning. This idea is common to many who have written about metaphor: it is found in the works of literary critics like Richards, Empson, and Winters; philosophers from Aristotle to Max Black; psychologists from Freud and earlier to Skinner and later; and linguists from Plato to Uriel Weinreich and George Lakoff. The idea takes many forms, from the relatively simple in Aristotle to the relatively complex in Black. The idea appears in writings which maintain that a literal paraphrase of a metaphor can be produced, but it is also shared by those who hold that typically no literal paraphrase can be found. Some stress the special insight metaphor can inspire and make much of the fact that ordinary language, in its usual functioning, yields no such insight. Yet this view too sees metaphor as a form of communication alongside ordinary communication; metaphor conveys truths or falsehoods about the world much as plainer language does, though the message may be considered more exotic, profound, or cunningly garbed.

The concept of metaphor as primarily a vehicle for conveying ideas, even if unusual ones, seems to me as wrong as the parent idea that a metaphor has a special meaning. I agree with the view that metaphors cannot be paraphrased, but I think this is not because metaphors say something too novel for literal expression but because there is nothing there to paraphrase. Paraphrase, whether possible or not, is appropriate to what is *said*: we try, in paraphrase, to say it another way. But if I am right, a metaphor doesn't say anything beyond its literal meaning (nor does its maker say anything, in using the metaphor, beyond the literal). This is not, of course, to deny that a metaphor has a point, nor that that point can be brought out by using further words.

In the past those who have denied that metaphor has a cognitive content in addition to the literal have often been out to show that metaphor is confusing, merely emotive, unsuited to serious, scientific, or philosophic discourse. My views should not be associated with this tradition. Metaphor is a legitimate device not only in literature but in science, philosophy, and the law; it is effective in praise and abuse, prayer and promotion, description and prescription. For the most part I don't disagree with Max Black, Paul Henle,

Nelson Goodman, Monroe Beardsley, and the rest in their accounts of what metaphor accomplishes, except that I think it accomplishes more and that what is additional is different in kind.

My disagreement is with the explanation of how metaphor works its wonders. To anticipate: I depend on the distinction between what words mean and what they are used to do. I think metaphor belongs exclusively to the domain of use. It is something brought off by the imaginative employment of words and sentences and depends entirely on the ordinary meanings of those words and hence on the ordinary meanings of the sentences they comprise.

It is no help in explaining how words work in metaphor to posit metaphorical or figurative meanings, or special kinds of poetic or metaphorical truth. These ideas don't explain metaphor, metaphor explains them. Once we understand a metaphor we can call what we grasp the 'metaphorical truth' and (up to a point) say what the 'metaphorical meaning' is. But simply to lodge this meaning in the metaphor is like explaining why a pill puts you to sleep by saying it has a dormative power. Literal meaning and literal truth conditions can be assigned to words and sentences apart from particular contexts of use. This is why adverting to them has genuine explanatory power.

I shall try to establish my negative views about what metaphors mean and introduce my limited positive claims by examining some false theories of the nature of metaphor.

A metaphor makes us attend to some likeness, often a novel or surprising likeness, between two or more things. This trite and true observations leads, or seems to lead, to a conclusion concerning the meaning of metaphors. Consider ordinary likeness or similarity: two roses are similar because they share the property of being a rose; two infants are similar by virtue of their infanthood. Or, more simply, roses are similar because each is a rose, infants, because each is an infant.

Suppose someone says 'Tolstoy was once an infant'. How is the infant Tolstoy like other infants? The answer comes pat: by virtue of exhibiting the property of infanthood, that is, leaving out some of the wind, by virtue of being an infant. If we tire of the phrase 'by virtue of', we can, it seems, be plainer still by saying the infant Tolstoy shares with other infants the fact that the predicate 'is an infant' applies to him; given the word 'infant', we have no trouble saying exactly how the infant Tolstoy resembles other infants. We could do it without the word 'infant'; all we need is other words that mean the same. The end result is the same. Ordinary similarity depends on groupings established by the ordinary meanings of words. Such similarity is natural and unsurprising to the extent that familiar ways of grouping objects are tied to usual meanings of usual words.

A famous critic said that Tolstoy was 'a great moralizing infant'. The Tolstoy referred to here is obviously not the infant Tolstoy but Tolstoy the adult writer; this is metaphor. Now in what sense is Tolstoy the writer similar to an infant? What we are to do, perhaps, is think of the class of objects which includes all ordinary infants and, in addition, the adult Tolstoy and then ask ourselves what special, surprising property the members of this class have in common. The appealing thought is that given patience we could come as close as need be to specifying the appropriate property. In any case, we could do the job perfectly if we found words that meant exactly what the metaphorical 'infant' means. The important point, from my perspective, is not whether we can find the perfect other words but the assumption that there is something to be attempted, a metaphorical meaning to be matched. So far I have been doing no more than crudely sketching how the concept of meaning may have crept into the analysis of metaphor, and the answer I have suggested is that since what we think of as garden variety similarity goes with what we think of as garden variety meanings, it is natural to posit unusual or metaphorical meanings to help explain the similarities metaphor promotes.

The idea, then, is that in metaphor certain words take on new, or what are often called 'extended', meanings. When we read, for example, that 'the Spirit of God moved upon the face of the waters', we are to regard the word 'face' as having an extended meaning (I disregard further metaphor in the passage). The extension applies, as it happens, to what philosophers call the extension of the word, that is, the class of entities to which it refers. Here the word 'face' applies to ordinary faces, and to waters in addition.

This account cannot, at any rate, be complete, for if in these contexts the words 'face' and 'infant' apply correctly to waters and to the adult Tolstoy, then waters really do have faces and Tolstoy literally was an infant, and all sense of metaphor evaporates. If we are to think of words in metaphors as directly going about their business of applying to what they propertly do apply to, there is no difference between metaphor and the introduction of a new term into our vocabulary: to make a metaphor is to murder it.

What has been left out is any appeal to the original meaning of the words. Whether or not metaphor depends on new or extended meanings, it certainly depends in some way on the original meanings; an adequate account of metaphor must allow that the primary or original meanings of words remain active in their metaphorical setting.

Perhaps, then, we can explain metaphor as a kind of ambiguity: in the context of a metaphor, certain words have either a new or an original meaning, and the force of the metaphor depends on our uncertainty as we waver between the two meanings. Thus when Melville writes that 'Christ was a chronometer', the effect of metaphor is produced by our taking 'chronometer' first in its ordinary sense and then in some extraordinary or metaphorical sense.

It is hard to see how this theory can be correct. For the ambiguity in the word, if there is any, is due to the fact that in ordinary contexts it means one thing and in the metaphorical context it means something else, but in the metaphorical context we do not necessarily hesitate over its meaning. When we do hesitate, it is usually to decide which of a number of metaphorical interpretations we shall accept; we are seldom in doubt that what we have is a metaphor. At any rate, the effectiveness of the metaphor easily outlasts the end of uncertainty over the interpretation of the metaphorical passage. Metaphor cannot, therefore, owe its effect to ambiguity of this sort.[2]

Another brand of ambiguity may appear to offer a better suggestion. Sometimes a word will, in a single context, bear two meanings where we are meant to remember and to use both. Or, if we think of wordhood as implying sameness of meaning, then we may describe the situation as one in which what appears as a single

[2] Nelson Goodman says metaphor and ambiguity differ chiefly 'in that the several uses of a merely ambiguous term are coeval and independent' while in metaphor 'a term with an extension established by habit is applied elsewhere under the influence of that habit'; he suggests that as our sense of the history of the 'two uses' in metaphor fades, the metaphorical word becomes merely ambiguous (*Languages of Art*, 71). In fact in many cases of ambiguity, one use springs from the other (as Goodman says) and so cannot be coeval. But the basic error, which Goodman shares with others, is the idea that two 'uses' are involved in metaphor in anything like the way they are in ambiguity.

word is in fact two. When Shakespeare's Cressida is welcomed bawdily into the Grecian camp, Nestor says, 'Our general doth salute you with a kiss.' Here we are to take 'general' two ways: once as applying to Agamemnon, who is the general; and once, since she is kissing everyone, as applying to no one in particular, but everyone in general. We really have a conjunction of two sentences: our general, Agamemnon, salutes you with a kiss; and everyone in general is saluting you with a kiss.

This is a legitimate device, a pun, but it is not the same device as metaphor. For in metaphor there is no essential need of reiteration; whatever meanings we assign the words, they keep through every correct reading of the passage.

A plausible modification of the last suggestion would be to consider the key word (or words) in a metaphor as having two different kinds of meaning at once, a literal and a figurative meaning. Imagine the literal meaning as latent, something that we are aware of, that can work on us without working in the context, while the figurative meaning carries the direct load. And finally, there must be a rule which connects the two meanings, for otherwise the explanation lapses into a form of the ambiguity theory. The rule, at least for many typical cases of metaphor, says that in its metaphorical role the word applies to everything that it applies to in its literal role, and then some.[3]

This theory may seem complex, but it is strikingly similar to what Frege proposed to account for the behaviour of referring terms in modal sentences and sentences about propositional attitudes like belief and desire. According to Frege, each referring term has two (or more) meanings, one which fixes its reference in ordinary contexts and another which fixes its reference in the special contexts created by modal operators or psychological verbs. The rule connecting the two meanings may be put like this: the meaning of the word in the special contexts makes the reference in those contexts to be identical with the meaning in ordinary contexts.

Here is the whole picture, putting Frege together with a Fregean view of metaphor: we are to think of a word as having, in addition to its mundane field of application or reference, two special or supermundane fields of application, one for metaphor and the other for modal contexts and the like. In both cases the original meaning

[3] The theory described is essentially that of Paul Henle, 'Metaphor'.

remains to do its work by virtue of a rule which relates the various meanings.

Having stressed the possible analogy between metaphorical meaning and the Fregean meanings for oblique contexts, I turn to an imposing difficulty in maintaining the analogy. You are entertaining a visitor from Saturn by trying to teach him to use the word 'floor'. You go through the familiar dodges, leading him from floor to floor, pointing and stamping and repeating the word. You prompt him to make experiments, tapping objects tentatively with his tentacle while rewarding his right and wrong tries. You want him to come out knowing not only that these particular objects or surfaces are floors but also how to tell a floor when one is in sight or touch. The skit you are putting on doesn't *tell* him what he needs to know, but with luck it helps him to learn it.

Should we call this process learning something about the world or learning something about language? An odd question, since what is learned is that a bit of language refers to a bit of the world. Still, it is easy to distinguish between the business of learning the meaning of a word and using the word once the meaning is learned. Comparing these two activities, it is natural to say that the first concerns learning something about language, while the second is typically learning something about the world. If your Saturnian has learned how to use the word 'floor', you may try telling him something new, that *here* is a floor. If he has mastered the word trick, you have told him something about the world.

You friend from Saturn now transports you through space to his home sphere, and looking back remotely at earth you say to him, nodding at the earth, 'floor'. Perhaps he will think this is still part of the lesson and assume that the word 'floor' applies properly to the earth, at least as seen from Saturn. But what if you thought he already knew the meaning of 'floor', and you were remembering how Dante, from a similar place in the heavens, saw the inhabited earth as 'the small round floor that makes us passionate'? Your purpose was metaphor, not drill in the use of language. What difference would it make to your friend which way he took it? With the theory of metaphor under consideration, very little difference, for according to that theory a word has a new meaning in a metaphorical context; the occasion of the metaphor would, therefore, be the occasion for learning the new meaning. We should agree that in some ways it makes relatively little difference whether, in a given context, we think a word is being used metaphorically or in a previously unknown, but literal way. Empson, in *Some Versions of Pastoral*, quotes these lines from Donne: 'As our blood labours to beget / Spirits, as like souls as it can, . . . / So must pure lover's soules descend. . . .' The modern reader is almost certain, Empson points out, to take the word 'spirits' in this passage metaphorically, as applying only by extension to something spiritual. But for Donne there was no metaphor. He writes in his *Sermons*, 'The Spirits . . . are the thin and active parts of the blood, and are a kind of middle nature, between soul and body.' Learning this does not matter much; Empson is right when he says, 'It is curious how the change in the word [that is, in what we think it means] leaves the poetry unaffected.'[4]

The change may be, in some cases at least, hard to appreciate, but unless there is a change, most of what is thought to be interesting about metaphor is lost. I have been making the point by contrasting learning a new use for an old word with using a word already understood; in one case, I said, our attention is directed to language, in the other, to what language is about. Metaphor, I suggested, belongs in the second category. This can also be seen by considering dead metaphors. Once upon a time, I suppose, rivers and bottles did not, as they do now, literally have mouths. Thinking of present usage, it doesn't matter whether we take the word 'mouth' to be ambiguous because it applies to entrances to rivers and openings of bottles as well as to animal apertures, or we think there is a single wide field of application that embraces both. What does matter is that when 'mouth' applied only metaphorically to bottles, the application made the hearer *notice* a likeness between animal and bottle openings. (Consider Homer's reference to wounds as mouths). Once one has the present use of the word, with literal application to bottles, there is nothing left to notice. There is no similarity to seek because it consists simply in being referred to by the same word.

Novelty is not the issue. In its context a word once taken for a metaphor remains a metaphor on the hundredth hearing, while a word may easily be appreciated in a new literal role on a first encounter. What we call the element of novelty or surprise in a

[4] W. Empson, *Some Versions of Pastoral*, 133.

metaphor is a built-in aesthetic feature we can experience again and again, like the surprise in Haydn's Symphony No. 94, or a familiar deceptive cadence.

If metaphor involved a second meaning, as ambiguity does, we might expect to be able to specify the special meaning of a word in a metaphorical setting by waiting until the metaphor dies. The figurative meaning of the living metaphor should be immortalized in the literal meaning of the dead. But although some philosophers have suggested this idea, it seems plainly wrong. 'He was burned up' is genuinely ambiguous (since it may be true in one sense and false in another), but although the slangish idiom is no doubt the corpse of a metaphor, 'He was burned up' now suggests no more than that he was very angry. When the metaphor was active, we would have pictured fire in the eyes or smoke coming out of the ears.

We can learn much about what metaphors mean by comparing them with similes, for a simile tells us, in part, what a metaphor merely nudges us into noting. Suppose Goneril had said, thinking of Lear, 'Old fools are like babes again'; then she would have used the words to assert a similarity between old fools and babes. What she did say, of course, was 'Old fools are babes again', thus using the words to intimate what the simile declared. Thinking along these lines may inspire another theory of the figurative or special meaning of metaphors: the figurative meaning of a metaphor is the literal meaning of the corresponding simile. Thus 'Christ was a chronometer' in its figurative sense is synonymous with 'Christ was like a chronometer', and the metaphorical meaning once locked up in 'He was burned up' is released in 'He was like someone who was burned up' (or perhaps 'He was like burned up').

There is, to be sure, the difficulty of identifying the simile that corresponds to a given metaphor. Virginia Woolf said that a highbrow is 'a man or woman of thoroughbred intelligence who rides his mind at a gallop across country in pursuit of an idea'. What simile corresponds? Something like this, perhaps: 'A highbrow is a man or woman whose intelligence is like a thoroughbred horse and who persists in thinking about an idea like a rider galloping across country in pursuit of . . . well, something.'

The view that the special meaning of a metaphor is identical with the literal meaning of a corresponding simile (however 'corresponding' is spelled out) should not be confused with the common theory that a metaphor is an elliptical simile.[5] This theory makes no distinction in meaning between a metaphor and some related simile and does not provide any ground for speaking of figurative, metaphorical, or special meanings. It is a theory that wins hands down so far as simplicity is concerned, but it also seems too simple to work. For if we make the literal meaning of the metaphor to be the literal meaning of a matching simile, we deny access to what we originally took to be the literal meaning of the metaphor, and we agreed almost from the start that *this* meaning was essential to the working of the metaphor, whatever else might have to be brought in in the way of a non-literal meaning.

Both the elliptical simile theory of metaphor and its more sophisticated variant, which equates the figurative meaning of the metaphor with the literal meaning of a simile, share a fatal defect. They make the hidden meaning of the metaphor all too obvious and accessible. In each case the hidden meaning is to be found simply by looking to the literal meaning of what is usually a painfully trivial simile. This is like that—Tolstoy like an infant, the earth like a floor. It is trivial because everything is like everything, and in endless ways. Metaphors are often very difficult to interpret and, so it is said, impossible to paraphrase. But with this theory, interpretation and paraphrase typically are ready to the hand of the most callow.

These simile theories have been found acceptable, I think, only because they have been confused with a quite different theory. Consider this remark by Max Black:

> When Schopenhauer called a geometrical proof a mousetrap, he was, according to such a view, *saying* (though not explicitly): 'A geometrical proof is *like* a mousetrap, since both offer a delusive reward, entice their victims by degrees, lead to disagreeable surprise, etc.' This is a view of metaphor as a condensed or elliptical *simile*.[6]

Here I discern two confusions. First, if metaphors are elliptical similes, they say *explicitly* what similes say, for ellipsis is a form of abbreviation, not of paraphrase or indirection. But, and this is the more important matter, Black's statement of what the metaphor says goes far beyond anything given by the corresponding simile. The simile simply says a geometrical proof is like a mousetrap. It no

[5] J. Middleton Murray says a metaphor is a 'compressed simile' (*Countries of the Mind*, 3). Max Black attributes a similar view to Alexander Bain, *English Composition and Rhetoric*.

[6] M. Black, 'Metaphor', 35.

more *tells* us what similarities we are to notice that the metaphor does. Black mentions three similarities, and of course we could go on adding to the list forever. But is this list, when revised and supplemented in the right way, supposed to give the *literal* meaning of the simile? Surely not, since the simile declared no more than the similarity. If the list is supposed to provide the figurative meaning of the simile, then we learn nothing about metaphor from the comparison with simile—only that both have the same figurative meaning. Nelson Goodman does indeed claim that 'the difference between simile and metaphor is negligible', and he continues, 'Whether the locution be "is like" or "is", the figure *likens* picture to person by picking out a certain common feature. . . .'[7] Goodman is considering the difference between saying a picture is sad and saying it is like a sad person. It is clearly true that both sayings liken picture to person, but it seems to me a mistake to claim that either way of talking 'picks out' a common feature. The simile says there is a likeness and leaves it to us to pick out some common feature or features; the metaphor does not explicitly assert a likeness, but if we accept it as a metaphor, we are again led to seek common features (not necessarily the same features the associated simile suggests; but that is another matter).

Just because a simile wears a declaration of similitude on its sleeve, it is, I think, far less plausible than in the case of metaphor to maintain that there is a hidden second meaning. In the case of simile, we note what it literally says, that two things resemble one another; we then regard the objects and consider what similarity would, in the context, be to the point. Having decided, we might then say the author of the simile intended us—that is, meant us—to notice that similarity. But having appreciated the difference between what the words meant and what the author accomplished by using those words, we should feel little temptation to explain what has happened by endowing the words themselves with a second, or figurative, meaning. The point of the concept of linguistic meaning is to explain what can be done with words. But the supposed figurative meaning of a simile explains nothing; it is not a feature of the word that the word has prior to and independent of the context of use, and it rests upon no linguistic customs except those that govern ordinary meaning.

[7] N. Goodman, *Languages of Art*, 77–8.

What words do do with their literal meaning in simile must be possible for them to do in metaphor. A metaphor directs attention to the same sorts of similarity, if not the same similarities, as the corresponding simile. But then the unexpected or subtle parallels and analogies it is the business of metaphor to promote need not depend, for their promotion, on more than the literal meanings of words.

Metaphor and simile are merely two among endless devices that serve to alert us to aspects of the world by inviting us to make comparisons. I quote a few stanzas of T. S. Eliot's 'The Hippopotamus':

> The broad-backed hippopotamus
> Rests on his belly in the mud;
> Although he seems so firm to us
> He is merely flesh and blood.
>
> Flesh and blood is weak and frail,
> Susceptible to nervous shock;
> While the True Church can never fail
> For it is based upon a rock.
>
> The hippo's feeble steps may err
> In compassing material ends,
> While the True Church need never stir
> To gather in its dividends.
>
> The 'potamus can never reach
> The mango on the mango-tree;
> But fruits of pomegranate and peach
> Refresh the Church from over sea.[8]

Here we are neither told that the Church resembles a hippopotamus (as in simile) not bullied into making this comparison (as in metaphor), but there can be no doubt the words are being used to direct our attention to similarities between the two. Nor should there be much inclination, in this case, to posit figurative meanings, for in what words or sentences would we lodge them? The hippopotamus really does rest on his belly in the mud; the True Church, the poem says literally, never can fail. The poem does, of course, intimate much that goes beyond the literal meaning of the words. But intimation is not meaning.

The argument so far has led to the conclusion that as much of metaphor as can be explained in terms of meaning may, and indeed

[8] T. S. Eliot, *Selected Poems*.

must, be explained by appeal to the literal meanings of words. A consequence is that the sentences in which metaphors occur are true or false in a normal, literal way, for if the words in them don't have special meanings, sentences don't have special truth. This is not to deny that there is such a thing as metaphorical truth, only to deny it of sentences. Metaphor does lead us to notice what might not otherwise be noticed, and there is no reason, I suppose, not to say these visions, thoughts, and feelings inspired by the metaphor are true or false.

If a sentence used metaphorically is true or false in the ordinary sense, then it is clear that it is usually false. The most obvious semantic difference between simile and metaphor is that all similes are true and most metaphors are false. The earth is like a floor, the Assyrian did come down like a wolf on the fold, because everything is like everything. But turn these sentences into metaphors, and you turn them false; the earth is like a floor, but it is not a floor; Tolstoy, grown up, was like an infant, but he wasn't one. We use a simile ordinarily only when we know the corresponding metaphor to be false. We say Mr S. is like a pig because we know he isn't one. If we had used a metaphor and said he was a pig, this would not be because we changed our mind about the facts but because we chose to get the idea across a different way.

What matters is not actual falsehood but that the sentence be taken to be false. Notice what happens when a sentence we use as a metaphor, believing it false, comes to be thought true because of a change in what is believed about the world. When it was reported that Hemingway's plane had been sighted, wrecked, in Africa, the New York *Mirror* ran a headline saying, 'Hemingway Lost in Africa', the word 'lost' being used to suggest he was dead. When it turned out he was alive, the *Mirror* left the headline to be taken literally. Or consider this case: a woman sees herself in a beautiful dress and says, 'What a dream of a dress!'—and then wakes up. The point of the metaphor is that the dress is like a dress one would dream of and therefore isn't a dream-dress. Henle provides a good example from *Anthony and Cleopatra* (2. 2):

> The barge she sat in, like a burnish'd throne
> Burn'd on the water

Here simile and metaphor interact strangely, but the metaphor would vanish if a literal conflagration were imagined. In much the same way the usual effect of a simile can be sabotaged by taking the comparison too earnestly. Woody Allen writes, 'The trial, which took place over the following weeks, was like a circus, although there was some difficulty getting the elephants into the courtroom'.[9]

Generally it is only when a sentence is taken to be false that we accept it as a metaphor and start to hunt out the hidden implication. It is probably for this reason that most metaphorical sentences are *patently* false, just as all similes are trivially true. Absurdity or contradiction in a metaphorical sentence guarantees we won't believe it and invites us, under proper circumstances, to take the sentence metaphorically.

Patent falsity is the usual case with metaphor, but on occasion patent truth will do as well. 'Business is business' is too obvious in its literal meaning to be taken as having been uttered to convey information, so we look for another use; Ted Cohen reminds us, in the same connection, that no man is an island.[10] The point is the same. The ordinary meaning in the context of use is odd enough to prompt us to disregard the question of literal truth.

Now let me raise a somewhat Platonic issue by comparing the making of a metaphor with telling a lie. The comparison is apt because lying, like making a metaphor, concerns not the meaning of words but their use. It is sometimes said that telling a lie entails what is false; but this is wrong. Telling a lie requires not that what you say be false but that you think it false. Since we usually believe true sentences and disbelieve false, most lies are falsehoods; but in any particular case this is an accident. The parallel between making a metaphor and telling a lie is emphasized by the fact that the same sentence can be used, with meaning unchanged, for either purpose. So a woman who believed in witches but did not think her neighbour a witch might say, 'She's a witch', meaning it metaphorically; the same woman, still believing the same of witches and her neighbour but intending to deceive, might use the same words to very different effect. Since sentence and meaning are the same in both cases, it is sometimes hard to prove which intention lay behind the saying of it; thus a man who says 'Lattimore's a Communist' and means to lie can always try to beg off by pleading a metaphor.

[9] Woody Allen, 'Condemned'.

[10] T. Cohen, 'Figurative Speech and Figurative Acts', 671. Since the negation of a metaphor seems always to be a potential metaphor, there may be as many platitudes among the potential metaphors as there are absurds among the actuals.

What makes the difference between a lie and a metaphor is not a difference in the words used or what they mean (in any strict sense of meaning) but in how the words are used. Using a sentence to tell a lie and using it to make a metaphor are, of course, totally different uses, so different that they do not interfere with one another, as say, acting and lying do. In lying, one must make an assertion so as to represent oneself as believing what one does not; in acting, assertion is excluded. Metaphor is careless of the difference. It can be an insult, and so be an assertion, to say to a man 'You are a pig'. But no metaphor was involved when (let us suppose) Odysseus addressed the same words to his companions in Circe's palace; a story, to be sure, and so no assertion—but the word, for once, was used literally of men.

No theory of metaphorical meaning or metaphorical truth can help explain how metaphor works. Metaphor runs on the same familiar linguistic tracks that the plainest sentences do; this we saw from considering simile. What distinguishes metaphor is not meaning but use—in this it is like assertion, hinting, lying, promising, or criticizing. And the special use to which we put language in metaphor is not—cannot be—to 'say something' special, no matter how indirectly. For a metaphor *says* only what shows on its face—usually a patent falsehood or an absurd truth. And this plain truth or falsehood needs no paraphrase—its meaning is given in the literal meaning of the words.

What are we to make, then, of the endless energy that has been, and is being, spent on methods and devices for drawing out the content of a metaphor? The psychologists Robert Verbrugge and Nancy McCarrell tell us that:

> Many metaphors draw attention to common systems of relationships or common transformations, in which the identity of the participants is secondary. For example, consider the sentences: *A car is like an animal*, *Tree trunks are straws for thirsty leaves and branches*. The first sentence directs attention to systems of relationships among energy consumption, respiration, self-induced motion, sensory systems, and, possibly a homunculus. In the second sentence, the resemblance is a more constrained type of transformation: suction of fluid through a vertically oriented cylindrical space from a source of fluid to a destination.[11]

Verbrugge and McCarrell don't believe there is any sharp line

[11] R. R. Verbrugge and N. S. McCarrell, 'Metaphoric Comprehension: Studies in Reminding and Resembling', 499.

between the literal and metaphorical uses of words; they think many words have a 'fuzzy' meaning that gets fixed, if fixed at all, by a context. But surely this fuzziness, however it is illustrated and explained, cannot erase the line between what a sentence literally means (given its context) and what it 'draws our attention to' (given its literal meaning as fixed by the context). The passage I have quoted is not employing such a distinction: what it says the sample sentences direct our attention to are facts expressed by paraphrases of the sentences. Verbrugge and McCarrell simply want to insist that a correct paraphrase may emphasize 'systems of relationships' rather than resemblances between objects.

According to Black's interaction theory, a metaphor makes us apply a 'system of commonplaces' associated with the metaphorical word to the subject of the metaphor: in 'Man is a wolf' we apply commonplace attributes (stereotypes) of the wolf to man. The metaphor, Black says, thus 'selects, emphasizes, suppresses, and organizes features of the principal subject by implying statements about it that normally apply to the subsidiary subject'.[12] If paraphrase fails, according to Black, it is not because the metaphor does not have a special cognitive content, but because the paraphrase 'will not have the same power to inform and enlighten as the original. . . . One of the points I most wish to stress is that the loss in such cases is a loss in cognitive content; the relevant weakness of the literal paraphrase is not that it may be tiresomely prolix or boringly explicit; it fails to be a translation because it fails to give the insight that the metaphor did.'[13]

How can this be right? If a metaphor has a special cognitive content, why should it be so difficult or impossible to set it out? If, as Owen Barfield claims, a metaphor 'says one thing and means another', why should it be that when we try to get explicit about what it means, the effect is so much weaker—'put it that way', Barfield says, 'and nearly all the tarning, and with it half the poetry, is lost.'[14] Why does Black think a literal paraphrase 'inevitably says too much—and with the wrong emphasis'? Why inevitably? Can't we, if we are clever enough, come as close as we please?

For that matter, how is it that a simile gets along without a special intermediate meaning? In general, critics do not suggest that a simile

[12] M. Black, 'Metaphor', 44–5. [13] Ibid., 46.
[14] O. Barfield, 'Poetic Diction and Legal Fiction', 55.

says one thing and means another—they do not suppose it *means* anything but what lies on the surface of the words. It may make us think deep thoughts, just as a metaphor does; how come, then, no one appeals to the 'special cognitive content' of the simile? And remember Eliot's hippopotamus; there there was neither simile nor metaphor, but what seemed to get done was just like what gets done by similes and metaphors, Does anyone suggest that the *words* in Eliot's poem have special meanings?

Finally, if words in metaphor bear a coded meaning, how can this meaning differ from the meaning those same words bear in the case where the metaphor *dies*—that is, when it comes to be part of the language? Why doesn't 'He was burned up' as now used and meant mean *exactly* what the fresh metaphor once meant? Yet all that the dead metaphor means is that he was very angry—a notion not very difficult to make explicit.

There is, then, a tension in the usual view of metaphor. For on the one hand, the usual view wants to hold that a metaphor does something no plain prose can possibly do and, on the other hand, it wants to explain what a metaphor does by appealing to a cognitive content—just the sort of thing plain prose is designed to express. As long as we are in this frame of mind, we must harbour the suspicion that it *can* be done, at least up to a point.

There is a simple way out of the impasse. We must give up the idea that a metaphor carries a message, that it has a content or meaning (except, of course, its literal meaning). The various theories we have been considering mistake their goal. Where they think they provide a method for deciphering an encoded content, they actually tell us (or try to tell us) something about the *effects* metaphors have on us. The common error is to fasten on the contents of the thoughts a metaphor provokes and to read these contents into the metaphor itself. No doubt metaphors often make us notice aspects of things we did not notice before; no doubt they bring surprising analogies and similarities to our attention; they do provide a kind of lens or lattice, as Black says, through which we view the relevant phenomena. The issue does not lie here but in the question of how the metaphor is related to what it makes us see.

It may be remarked with justice that the claim that a metaphor provokes or invites a certain view of its subject rather than saying it straight out is a commonplace; so it is. Thus Aristotle says metaphor leads to a 'perception of resemblances'. Black, following Richards, says a metaphor 'evokes' a certain response: 'a suitable hearer will be led by a metaphor to construct a . . . system.'[15] This view is neatly summed up by what Heracleitus said of the Delphic oracle: 'It does not say and it does not hide, it intimates.'[16]

I have no quarrel with these descriptions of the effects of metaphor, only with the associated views as to *how* metaphor is supposed to produce them. What I deny is that metaphor does its work by having a special meaning, a specific cognitive content. I do not think, as Richards does, that metaphor produces its result by having a meaning which results from the interaction of two ideas; it is wrong, in my view, to say, with Owen Barfield, that a metaphor 'says one thing and means another'; or with Black that a metaphor asserts or implies certain complex things by dint of a special meaning and *thus* accomplishes its job of yielding an 'insight'. A metaphor does its work through other intermediaries—to suppose it can be effective only by conveying a coded message is like thinking a joke or a dream makes some statement which a clever interpreter can restate in plain prose. Joke or dream or metaphor can, like a picture or a bump on the head, make us appreciate some fact—but not by standing for, or expressing, the fact.

If this is right, what we attempt in 'paraphrasing' a metaphor cannot be to give its meaning, for that lies on the surface; rather we attempt to evoke what the metaphor brings to our attention. I can imagine someone granting this and shrugging it off as no more than an insistence on restraint in using the word 'meaning'. This would be wrong. The central error about metaphor is most easily attacked when it takes the form of a theory of metaphorical meaning, but behind that theory, and statable independently, is the thesis that associated with a metaphor is a definite cognitive content that its author wishes to convey and that the interpreter must grasp if he is to get the message. This theory is false as a full account of metaphor, whether or not we call the purported cognitive content a meaning.

It should make us suspect the theory that it is so hard to decide, even in the case of the simplest metaphors, exactly what the content is supposed to be. The reason it is often so hard to decide is, I think, that we imagine there is a content to be captured when all the while we are in fact focusing on what the metaphor makes us notice. If

[15] M. Black, 'Metaphor', 41.

[16] I use Hannah Arendt's attractive translation of 'σημαίνει'; it clearly should not be rendered as 'means' in this context.

what the metaphor makes us notice were finite in scope and propositional in nature, this would not in itself make trouble; we would simply project the content the metaphor brought to mind on to the metaphor. But in fact there is no limit to what a metaphor calls to our attention, and much of what we are caused to notice is not propositional in character. When we try to say what a metaphor 'means', we soon realize there is no end to what we want to mention.[17] If someone draws his finger along a coastline on a map, or mentions the beauty and deftness of a line in a Picasso etching, how many things are drawn to your attention? You might list a great many, but you could not finish since the idea of finishing would have no clear application. How many facts or propositions are conveyed by a photograph? None, an infinity, or one great unstatable fact? Bad question. A picture is not worth a thousand words, or any other number. Words are the wrong currency to exchange for a picture.

It's not only that we can't provide an exhaustive catalogue of what has been attended to when we are led to see something in a new light; the difficulty is more fundamental. What we notice or see is not, in general, propositional in character. Of course it *may* be, and when it is, it usually may be stated in fairly plain words. But if I show you Wittgenstein's duck-rabbit, and I say, 'It's a duck', then with luck you see it as a duck; if I say, 'It's a rabbit', you see it as a rabbit. But no proposition expresses what I have led you to see. Perhaps you have come to realize that the drawing can be seen as a duck or as a rabbit. But one could come to know this without ever seeing the drawing as a duck or as a rabbit. Seeing as is not seeing that. Metaphor makes us see one thing as another by making some literal statement that inspires or prompts the insight. Since in most cases what the metaphor prompts or inspires is not entirely, or even at all, recognition of some truth or fact, the attempt to give literal expression to the content of the metaphor is simply misguided.

The theorist who tries to explain a metaphor by appealing to a

[17] Stanley Cavell mentions the fact that most attempts at paraphrase end with 'and so on' and refers to Empson's remark that metaphors are 'pregnant' ('Aesthetic Problems of Modern Philosophy', 79). But Cavell doesn't explain the endlessness of paraphrase as I do, as can be learned from the fact that he thinks it distinguishes metaphor from some ('but perhaps not all') literal discourse. I hold that the endless character of what we call the paraphrase of a metaphor springs from the fact that it attempts to spell out what the metaphor makes us notice, and to this there is no clear end. I would say the same for any use of language.

hidden message, like the critic who attempts to state the message, is then fundamentally confused. No such explanation or statement can be forthcoming because no such message exists.

Not, of course, that interpretation and elucidation of a metaphor are not in order. Many of us need help if we are to see what the author of a metaphor wanted us to see and what a more sensitive or educated reader grasps. The legitimate function of so-called paraphrase is to make the lazy or ignorant reader have a vision like that of the skilled critic. The critic is, so to speak, in benign competition with the metaphor maker. The critic tries to make his own art easier or more transparent in some respects that the original, but at the same time he tries to reproduce in others some of the effects the original had on him. In doing this the critic also, and perhaps by the best method at his command, calls attention to the beauty or aptness, the hidden power, of the metaphor itself.

18 *Communication and Convention*

Convention figures conspicuously in many of our activities, for example in playing tarot, in speaking, and in eating. In playing tarot, convention is essential, in eating it is not. In explaining what it is to play tarot we could not leave out of account the rules that define the game; in explaining what it is to eat no mention of rules or conventions needs to be made. What is the case with speech? Are conventions mere conveniences or social flourishes, or are they necessary to the existence of communication by language?

The question is delicate because it concerns not the truth of the claim that speech is convention-bound, but the importance and role of convention in speech. The issue may be put counterfactually: could there be communication by language without convention? According to David Lewis, 'It is a platitude—something only a philosopher would dream of denying—that there are conventions of language.'[1] Certainly it would be absurd to deny that many conventions *involve* speech, such as saying 'Good morning' no matter what the weather is like; but this is not the sort of convention on which the existence of language depends. No doubt what Lewis has in mind is the idea that the connection between words and what they mean is conventional. And perhaps only a philosopher would deny this; but if so, the reason may be that only a philosopher would say it in the first place. What *is* obvious enough to be a platitude is that the use of a particular sound to refer to, or mean, what it does is *arbitrary*. But while what is conventional is in some sense arbitrary, what is arbitrary is not necessarily conventional.

In one respect we describe a language completely when we say

[1] D. Lewis, 'Languages and Language', 7.

what counts as a meaningful utterance and what each actual or potential utterance means. But such descriptions assume we already know what it is for an utterance to have a particular meaning. Light on *this* question—the traditional problem of meaning—requires us to connect the notion of meaning with beliefs, desires, intentions, and purposes in an illuminating way. It is mainly in making the connection, or connections, between linguistic meaning and human attitudes and acts described in non-linguistic terms that convention is asked to do its work. And here there are many different theories that have been proposed. I shall divide them into three kinds: first, there are theories that claim there is a convention connecting sentences in one or another grammatical mood (or containing an explicit performative phrase) with illocutionary intentions, or some broader purpose; second, there are theories that look to a conventional use for each sentence; and third, there are theories to the effect that there is a convention that ties individual words to an extension or intension. These are not competing theories. Depending on details, all combinations of these theories are possible. I discuss the three sorts of theory in the order just listed.

In an early, and influential, article Michael Dummett maintained that there is a convention that governs our use of declarative sentences.[2] As he has put it more recently:

> ... the utterance of a [declarative] sentence does not need a particular context to give it a point ... The utterance of a sentence serves to assert something ... there is a general convention whereby the utterance of a sentence, except in special contexts, is understood as being carried out with the intention of uttering a true sentence.[3]

This is a complex, and perhaps not entirely transparent, dictum, but I interpret it as follows. There is a conventional connection between uttering a declarative sentence and using it to make an assertion (one is making an assertion except in special contexts); and there is a conceptual (and perhaps conventional) connection between making an assertion and the intention to say what is true. The plausibility of this interpretation is brought out, I think, by Dummett's most convincing argument. He begins by examining Tarski-style truth definitions. Dummett reminds us (following, though he probably did not know this, an earlier paper by Max Black[4]) that while Tarski

[2] M. Dummett, 'Truth'.
[3] M. Dummett, *Frege: Philosophy of Language*, 298.
[4] M. Black, 'The Semantic Definition of Truth'.

showed how, in principle, to construct a truth definition for particular (formalized) languages, he did not, and indeed proved that no one could, define truth in general, at least using his method. Tarski was therefore not able to say what it was that made each definition of truth a definition of the same concept. Convention T, to which Tarski appealed for a criterion of the correctness of a truth definition, does not specify what truth in general is, but makes use of the intuitive grasp we have of the concept.

Dummett drew an analogy between truth and the concept of winning at a game. If we want to know what winning at a game is, we will not be satisfied by being told the definition of winning for each of several games; we want to know what makes the situation defined for each game a case of winning. Thinking of truth, the problem could be put this way: if we were exposed to speakers of a language we did not know, and were given a Tarski-style truth definition, how could we tell whether the definition applied to that language? A good question; but I do not believe it can be answered by attending to Dummett's proposed convention. For it seems to me that nothing in language corresponds in relevant ways to winning in a game. The point is important because if Dummett is right, to understand what it is in language that is like winning in a game is to make the crucial connection between meaning as described in a theory of truth and the use of language in contexts of communication.[5]

Winning in a game like chess has these characteristics: first, people who play usually want to win. Whether they want to win or not, it is a condition of playing that they *represent* themselves as wanting to win. This is not the same as pretending they want to win, or trying to get others to believe they want to win. But perhaps representing oneself as wanting to win does entail that one can be reproached if it is found that he does not want to win or isn't trying to win. Second, one can win only by making moves defined by the rules of the game, and winning is wholly defined by the rules. Finally, winning can be, and often is, an end in itself.[6] As far as I can see, no linguistic behaviour has this combination of features; if so,

[5] It is not pertinent to my argument that Dummett does not believe a theory of truth can serve as a theory of meaning. The issue here is whether or not there is a convention of a certain sort governing our utterances of (declarative) sentences.

[6] The distinction between activities that can be ends in themselves, such as playing the flute, and those that serve some further end, such as building a house, comes, of course, from Aristotle: *Nichomachean Ethics* 1094a; *Magna Moralia* 1211b.

Dummett's analogy between games and language is radically defective.

Is speaking the truth, in the sense of intentionally uttering a sentence that happens to be true, like winning? It is in this respect, that what it is to speak the truth is what a theory of truth seeks to define. *In so far, then*, as the truth conditions of utterances are known to speakers and interpreters in advance, and agreed upon as a condition of communication, speaking the truth has one of the features of winning. (I'll question how far this is true later.) But it lacks the others, for people who utter a sentence do not usually want to speak true sentences. Sometimes they do, and very often they don't. Nor, in order to play the speech game, do they have to represent themselves as intending or wanting to speak the truth; there is no general presumption that someone who utters a declarative sentence wants or intends to speak the truth, nor that, if he does, he does it intentionally. Finally, speaking the truth, in the sense of uttering a true sentence, is never an end in itself.

Assertion, in contrast to speaking the truth, may seem a likelier candidate for linguistic counterpart of winning. Someone who makes an assertion represents himself as believing what he says, and perhaps as being justified in his belief. And since we want our beliefs to be true, it seems right to agree with Dummett that when someone makes an assertion, he represents himself as intending to say what is true. (This is how I take Dummett's remark that the speaker is 'understood' to have the intention of uttering a true sentence.) As in playing a game, the representation may or may not be deceitful. (The liar makes an assertion.) The asserter may or may not, in making his assertion, intend to cause his hearer to believe he believes what he says. Making an assertion is, then, like playing a game in a respect in which speaking the truth is not: there is a public presumption of purpose. In other respects, however, assertion is unlike winning, for what constitutes the making of an assertion is not governed by agreed rules or conventions.[7]

If the concept of assertion is to provide a conventional bridge between purpose and truth, two things must hold: there must be conventions governing assertion, and there must be a convention linking assertion to what is believed true. I think neither of these claims holds.

[7] On this point I am much indebted to Sue Larson.

Many philosophers have thought there were conventions governing assertion. Thus Dummett, in a phrase I omitted from an earlier quotation, says, 'The utterance of a sentence serves to assert something . . .'[8] Let us first consider whether assertion is governed by conventions. Of course, if it is a convention that a sentence means what it literally does when uttered, then convention is involved in all utterances and hence in assertion. But literal meaning may not (and in my view does not) go beyond truth conditions. And no one will deny, I suppose, that the same declarative sentence may have the same meaning when used to make an assertion, to tell a joke, to annoy a bore, to complete a rhyme, or to ask a question. So if there is a convention, it must be further conventional trappings of the utterance that make it an assertion. It is not enough, of course, to say that *something* in the context makes it an assertion. This is true, but proves nothing about convention. And we may be able to say what it is in the context that makes it an assertion, though in fact I think we can say only some rather vague and incomplete things. But even if the necessary and sufficient conditions were explicit and agreed upon by all hands it would not yet follow that the conditions were conventional. We all agree that a horse must have four legs, but it is not a convention that horses have four legs.

There is something more about assertion that suggests that convention may be involved, and this is the fact that in making an assertion, the asserter must intend to make an assertion, and he must intend that this intention be recognized by his audience. Assertions are intended to be public performances where the clues are adequate to identify the character of the performance as assertoric. So it is natural to think it would be useful if there were a convention, as a convenience in making our assertive intentions clear.

But Frege was surely right when he said, 'There is no word or sign in language whose function is simply to assert something.' Frege, as we know, set out to rectify matters by inventing such a sign, the turnstile '⊢'. And here Frege was operating on the basis of a sound principle: if there is a conventional feature of language, it can be made manifest in the symbolism. However, before Frege invented the assertion sign he ought to have asked himself why no such sign existed before. Imagine this: the actor is acting a scene in which there is supposed to be a fire. (Albee's *Tiny Alice*, for example.) It is

[8] See footnote 3.

his role to imitate as persuasively as he can a man who is trying to warn others of a fire. 'Fire!' he screams. And perhaps he adds, at the behest of the author, 'I mean it! Look at the smoke!' etc. And now a real fire breaks out, and the actor tries vainly to warn the real audience. 'Fire!' he screams, 'I mean it! Look at the smoke!' etc. If only he had Frege's assertion sign.

It should be obvious that the assertion sign would do no good, for the actor would have used it in the first place, when he was only acting. Similar reasoning should convince us that it is no help to say that the stage, or the proscenium arch, creates a conventional setting which negates the convention of assertion. For if that were so, the acting convention could be put into symbols also; and of course no actor or director would use it. The plight of the actor is always with us. There is no known, agreed upon, publicly recognizable convention for making assertions. Or, for that matter, giving orders, asking questions, or making promises. These are all things we do, often successfully, and our success depends in part on our having made public our intention to do them. But it was not thanks to a convention that we succeeded.

The second point of Dummett's claim is that there is a convention that in making an assertion a speaker is 'understood' to be speaking with 'the intention of uttering a true sentence'. This also seems to me to be wrong, though in a somewhat different way. What is understood is that the speaker, if he has asserted something, has represented himself as believing it—as uttering a sentence he believes true, then. But this is not a convention, it is merely part of the analysis of what assertion is. To assert is, among other things, to represent oneself as believing what one asserts. It is clear that there cannot be a conventional sign that shows that one is saying what one believes; for every liar would use it. Convention cannot connect what may always be secret—the intention to say what is true—with what must be public—making an assertion. There is no convention of sincerity.

If literal meaning is conventional, then the difference in the grammatical moods—declarative, imperative, interrogative, optative—is conventional. These differences are in the open and intended to be recognized; syntax alone usually does the job. What this shows is that grammatical mood and illocutionary force, no matter how closely related, cannot be related simply by convention.

Although I have concentrated on assertion, similar considerations apply to illocutionary forces of all kinds. My main interest here, however, is not in the nature of illocutionary force, or in such acts as asserting, promising, and commanding, but in the idea that convention can link what our words mean—their literal semantic properties, including truth—and our purposes in using them, for example, to speak the truth.

We have been discussing claims that there are comprehensive purposes tied by convention to the enterprise of linguistic communication—purposes that, in Dummett's word, give us the 'point' of using language. I turn now to theories of quite a different sort, that attempt to derive the literal meanings of entire sentences (not just the mood indicators) from the non-linguistic purposes their utterances serve. I am concerned in the present essay with theories which make the derivation depend on convention.

Stated crudely, such theories maintain that there is a single use (or some finite number of uses) to which a given sentence is tied, and this use gives the meaning of the sentence. Since in fact there are endless uses to which a sentence, with meaning unchanged, can be put, the connection between a single use (or finite number of uses) and the sentence is conventional; it is a use that can be called standard.

This is too simple, of course, but it is an appealing and natural idea. For there does seem to be an important connection between a sentence like 'Eat your eggplant' and the intention, in uttering this sentence, to get someone to eat his eggplant. Getting someone to eat his eggplant is, you might say, what the English sentence 'Eat your eggplant' was made to do. If this intuition could be explicitly stated and defended in a non-question-begging way, there would be promise of an account of literal meaning in terms of the ordinary non-linguistic purposes that always lie behind the utterances of sentences.

There are intentions embedded in all linguistic utterances such that if we could detect them we would usually know what the words uttered literally meant. For someone cannot utter the sentence 'Eat the eggplant' with the words literally meaning that someone is to eat his eggplant unless he intends the sentence to have that meaning, and intends his audience to interpret it as having that meaning. Of course the mere intention does not *give* the sentence that meaning; but if it is uttered with the intention of uttering a sentence with that meaning, and it does not in fact have that meaning, then it has no linguistic meaning at all. Literal meaning and intended literal meaning must coincide if there is to be a literal meaning. But this fact, while true and important, is of no direct help in understanding the concept of literal meaning, since the crucial intention must be characterized by reference to the literal meaning. Nor can convention make a contribution here, for we were looking to convention to convert non-linguistic purposes into performances with a literal meaning. A convention that connected the intention to use words with a certain literal meaning with the literal meaning of those words would not explain the concept of literal meaning, but would depend upon it.

What we seek are intentions characterized in non-linguistic terms—*ulterior purposes* in uttering sentences. (This concept may be related to what Austin called perlocutionary acts.)

I mentioned briefly before, and now want to insist on, the fact that linguistic utterances always have an ulterior purpose; this was one of the reasons I gave for claiming that no purely linguistic activity is like winning at a game. There is perhaps some element of stipulation here, but I would not call it a linguistic act if one spoke 'words' merely to hear the sounds, or to put someone to sleep; an action counts as linguistic only if literal meaning is relevant. But where meaning is relevant, there is always an ulterior purpose. When one speaks, one aims to instruct, impress, amuse, insult, persuade, warn, remind, or aid a calculation. One may even speak with the intention of boring an audience; but not by hoping no one will attend to the meaning.

If I am right that each use of language has an ulterior purpose, then one must always intend to produce some non-linguistic effect through having one's words interpreted. Max Black has denied this, pointing out that '. . . a man may outline a lecture, or write a note to remind himself of an appointment, or simply utter certain words, such as "What a lovely day!" in the absence of an audience.'[9] The first two cases here are clearly cases where the meaning matters, and there is an audience which is intended to interpret the words: oneself at a later time. In the last case it would be tendentious to insist that one is speaking to oneself; yet it matters what words are used, what they mean. And there must be some *reason* for using those words, with their meaning, rather than others. Black quotes Chomsky to similar effect:

[9] M. Black, 'Meaning and Intention: An Examination of Grice's Views', 264.

Though consideration of intended effects avoids some problems, it will at best provide an analysis of successful communication, but not of meaning or the use of language, which need not involve communication, or even the attempt to communicate. If I use language to express or clarify my thoughts, or with the intent to deceive, to avoid an embarrassing silence, or in a dozen other ways, my words have a strict meaning and I can very well mean what I say, but the fullest understanding of what I intend my audience (if any) to believe or do might give little or no indication of the meaning of my discourse.[10]

In this passage it seems to me Chomsky arrives at a correct conclusion from confused or irrelevant premises. The issue is whether or not the meanings of sentences can be derived from the non-linguistic intentions of a speaker. Chomsky concludes, correctly, I believe, that they cannot. But it is irrelevant to the conclusion whether intended effects must involve someone other than the speaker, and unimportant to the argument whether there are intended effects. Speaking or writing in order to clarify thoughts certainly posits an intended effect. Nor does it matter just how we use the word 'communicate'. What matters is whether an activity is interestingly considered linguistic when meanings are not intended to be put to use. Lying is a case where literal meaning is essential; the liar has an ulterior purpose that is served only if his words are understood as having the meaning he intends.

Where Chomsky is right, as I said, is in claiming that no amount of knowledge of what I intend my audience to believe or do will necessarily yield the literal meaning of my utterance. Even this claim must, as we have seen, be limited to a description of my intention in non-linguistic terms. For if I intend to get my audience to do or believe something, it must be through their correct interpretation of the literal meaning of my words.

It is now relatively clear what a convention must do if it is to relate non-linguistic purposes in uttering sentences—ulterior purposes—with the literal meanings of those sentences when uttered. The convention must pick out, in a way understood by both speaker and hearer, and in an intentionally identifiable way, those cases in which the ulterior purpose directly yields the literal meaning. I mean, for example, a case where, in uttering the words 'Eat your eggplant' with their normal meaning in English, a speaker intends to get a hearer to eat his eggplant through the hearer's understanding of the

[10] N. Chomsky, *Problems of Knowledge and Freedom*, 19.

words and the illocutionary force of the utterance. And here, once again, it seems to me not only that there is no such convention, but that there cannot be. For even if, contrary to what I have argued, some convention governs the illocutionary force of the utterance, the connection with the intention that the request or order be carried out would require that the speaker be *sincere*—that what he represents himself as wanting or trying to do he in fact wants or is trying to do. But nothing is more obvious than that there cannot be a convention that signals sincerity.

It is no help, I must repeat, to say that the convention is that the sentence always means what the ulterior purpose would reveal if the speaker were sincere, serious, etc. This is at best a partial *analysis* of the relations between literal meaning, sincerity, and intention. It does not suggest publicly recognized tests, criteria, or practices.

Sometimes it is suggested that a language could never be learned except in an atmosphere of honest assertions (commands, promises, etc.). Even if this were true, it would prove nothing about a supposed role for convention. But I am also sceptical about the claim itself, partly because so much language learning takes place during games, in hearing stories, and in pretence, and partly because the acquisition of language cannot to such an extent depend on our luck in having truthful, sober, assertive playmates and parents.

It is in the nature of a game like chess or tarot not only that there are mutually agreed criteria of what it is to play, but mutually agreed criteria of what it is to win. It is essential to these games that normally there is no question about the outcome. And it is also in the nature of such games that winning can be an end in itself, and that players represent themselves as wanting or trying to win. But the criteria for deciding what an utterance literally means, given by a theory of truth or meaning for the speaker, do not decide whether he has accomplished his ulterior purpose, nor is there any general rule that speakers represent themselves as having any further end than that of using words with a certain meaning and force. The ulterior purpose may or may not be evident, and it may or may not help an interpreter determine the literal meaning. I conclude that it is not an accidental feature of language that the ulterior purpose of an utterance and its literal meaning are independent, in the sense that the latter cannot be derived from the former: it is of the essence of language. I call this feature of language the principle of *the autonomy of meaning*. We came across an application when discussing illocutionary force, where it took the form of the discovery that what is put into the literal meaning then becomes available for any ulterior (non-linguistic) purpose—and even any illocutionary performance.[11]

Before leaving the discussion of theories of the first two kinds, the following remark may be useful. Nothing I have said has been intended to show there is no connection between the mood indicators and the *idea* of a certain illocutionary act. I believe there is such a connection. An utterance of an imperative sentence, for example, quite literally labels itself as an act of ordering. But this is just part of the literal meaning of the uttered words, and establishes no relation, conventional or otherwise, between the illocutionary intentions of the speaker and his words. It is easy to confuse two quite different theses: on the one hand, the (correct) thesis that every utterance of an imperative *labels* itself (truly or falsely) an order, and the thesis that there is a convention that under 'standard' conditions the utterance of an imperative *is* an order. The first thesis does, while the second thesis can't, explain the difference in meaning between an imperative and a declarative sentence, a difference which exists quite independently of illocutionary force. The second thesis can't explain this because it postulates a convention that is in force only under 'standard' conditions. You can't *use* a convention by breaking it, you can only *abuse* it. But the difference between an imperative and a declarative can, and very often is, quite properly used in situations where mood and illocutionary force are not 'standard'. It will not help to point to such cases as an actor wearing a crown to indicate that he is playing the part of a king. If a convention is involved here, it is a convention governing literal meaning. Wearing a crown, whether done in jest or in earnest, is just like saying 'I am the king'.

The gist of this remark applies also to the kind of theory that tries to derive the literal meaning of each sentence from a 'standard' use. Since the literal meaning operates as well when the use is absent as when it is present, no convention that operates only in 'standard' situations can give the literal meaning.

We have considered the idea that linguistic activity in general is like a game in that there is a conventional purpose (saying what is true, winning) which can be achieved only by following, or using, agreed and public rules. Then we discussed the claim that the literal

[11] For further discussion, and a suggestion as to how illocutionary force and grammatical mood are related, see Essay 8.

meaning of each sentence is related by a convention to a standard non-linguistic end (an ulterior purpose). Both of these views turned out, on examination, to be untenable. Now it is time to evaluate the 'platitude' that the meaning of a word is conventional, that is, that it is a convention that we assign the meaning we do to individual words and sentences when they are uttered or written.

According to David Lewis[12] a convention is a *regularity* R in action, or action and belief, a regularity in which more than one person must be involved. The regularity has these properties:

(1) Everyone involved conforms to R and (2) believes that others also conform. (3) The belief that others conform to R gives all involved a good reason to conform to R. (4) All concerned prefer that there should be conformity to R. (5) R is not the only possible regularity meeting the last two conditions. (6) Finally, everyone involved knows (1)–(5) and knows that everyone else knows (1)–(5), etc.

Tyler Burge has raised reasonable doubts about parts of the last condition (does a convention require that everyone know there are alternatives?)[13] and I have misgivings myself. But here there is no reason to debate the details of the analysis of the concept of convention; what is relevant is to raise a question whether, or to what extent, convention in any reasonable sense helps us understand linguistic communication. So instead of asking, for example, what 'conforming' to a regularity adds to the regularity itself, I shall simply grant that something like Lewis's six conditions does hold roughly for what we call speakers of the same language. How fundamental a fact is this about language?

Lewis's analysis clearly requires that there be at least two people involved, since convention depends on a mutually understood practice. But nothing in the analysis requires more than two people. Two people could have conventions, and could share a language.

What exactly is the necessary convention? It cannot be that speaker and hearer mean the same thing by uttering the same sentences. For such conformity, while perhaps fairly common, is not necessary to communication. Each speaker may speak his different language, and this will not hinder communication as long as each hearer understands the one who speaks. It could even happen that

[12] D. Lewis, 'Languages and Language', 5, 6.
[13] T. Burge, 'Reasoning about Reasoning'.

every speaker from the start had his own quite unique way of speaking. Something approaching this is in fact the case, of course. Different speakers have different stocks of proper names, different vocabularies, and attach somewhat different meanings to words. In some cases this reduces the level of mutual understanding; but not necessarily, for as interpreters we are very good at arriving at a correct interpretation of words we have not heard before, or of words we have not heard before with meanings a speaker is giving them.

Communication does not demand, then, that speaker and hearer mean the same thing by the same words; yet convention requires conformity on the part of at least two people. However, there remains a further form of agreement that is necessary: if communication succeeds, speaker and hearer must assign the same meaning to the speaker's words. Further, as we have seen, the speaker must intend the hearer to interpret his words in the way the speaker intends, and he must have adequate reason to believe that the hearer will succeed in interpreting him as he intends. Both speaker and hearer must believe the speaker speaks with this intention, and so forth; in short, many of Lewis's conditions would seem to be satisfied. It's true that this is at best an attenuated sense of a practice or convention, remote from the usual idea of a common practice. Still, once might insist that this much of a mutally understood method of interpretation is the essential conventional core in linguistic communication.

But the most important feature of Lewis's analysis of convention—regularity—has yet to be accounted for. Regularity in this context must mean regularity over time, not mere agreement at a moment. If there is to be a convention in Lewis's sense (or in any sense, I would say), then something must be seen to repeat or recur over time. The only candidate for recurrence we have is the interpretation of sound patterns: speaker and hearer must repeatedly, intentionally, and with mutual agreement, interpret relevantly similar sound patterns of the speaker in the same way (or ways related by rules that can be made explicit in advance).

I do not doubt that all human linguistic communication does show a degree of such regularity, and perhaps some will feel inclined to make it a condition of calling an activity linguistic that there should be such regularity. I have doubts, however, both about the clarity of the claim and its importance in explaining and describing communication. The clarity comes into question because it is very difficult to say exactly how speaker's and hearer's theories for interpreting the speaker's words must coincide. They must, of course, coincide *after* an utterance has been made, or communication is impaired. But unless they coincide in advance, the concepts of regularity and convention have no definite purchase. Yet agreement on what a speaker means by what he says can surely be achieved even though speaker and hearer have different advance theories as to how to interpret the speaker. The reason this can be is that the speaker may well provide adequate clues, in what he says, and how and where he says it, to allow a hearer to arrive at a correct interpretation. Of course the speaker must have *some* idea how the hearer is apt to make use of the relevant clues; and the hearer must know a great deal about what to expect. But such general knowledge is hard to reduce to rules, much less conventions or practices.

It is easy to misconceive the role of society in language. Language is, to be sure, a social art. But it is an error to suppose we have seen deeply into the heart of linguistic communication when we have noticed how society bends linguistic habits to a public norm. What is conventional about language, if anything is, is that people tend to speak much as their neighbours do. But in indicating this element of the conventional, or of the conditioning process that makes speakers rough linguistic facsimiles of their friends and parents, we explain no more than the convergence; we throw no light on the essential nature of the skills that are thus made to converge.

This is not to deny the practical, as contrasted with the theoretical, importance of social conditioning. What common conditioning ensures is that we may, up to a point, assume that the same method of interpretation that we use for others, or that we assume others use for us, will work for a new speaker. We do not have the time, patience, or opportunity to evolve a new theory of interpretation for each speaker, and what saves us is that from the moment someone unknown to us opens his mouth, we know an enormous amount about the sort of theory that will work for him—or we know we know no such theory. But if his first words are, as we say, English, we are justified in assuming he has been exposed to linguistic conditioning similar to ours (we may even guess or know differences). To buy a pipe, order a meal, or direct a taxi driver, we go on this assumption. Until proven wrong; at which point we can revise our theory of what he means on the spur of the moment. The longer talk continues the better our theory becomes, and the more finely adapted to the individual speaker. Knowledge of the conventions of language is thus a practical crutch to interpretation, a crutch we cannot in practice afford to do without—but a crutch which, under optimum conditions for communication, we can in the end throw away, and could in theory have done without from the start.

The fact that radical interpretation is so commonplace—the fact, that is, that we use our standard method of interpretation only as a useful starting point in understanding a speaker—is hidden from us by many things, foremost among them being that syntax is so much more social than semantics. The reason for this, roughly stated, is that what forms the skeleton of what we call a language is the pattern of inference and structure created by the logical constants: the sentential connectives, quantifiers, and devices for cross-reference. If we can apply our general method of interpretation to a speaker at all—if we can make even a start in understanding him on the assumption that his language is like ours, it will thus be because we can treat his structure-forming devices as we treat ours. This fixes the logical form of his sentences, and determines the parts of speech. No doubt some stock of important predicates must translate in the obvious homophonic way if we are to get far fast; but we can then do very well in interpreting, or reinterpreting new, or apparently familiar, further predicates.

This picture of how interpretation takes place puts the application of formal methods to natural language in a new light. It helps show why formal methods are at their best applied to syntax; here at least there is good reason to expect the same model to fit a number of speakers fairly well. And there is no clear reason why each hypothesized method of interpretation should not be a formal semantics for what we may in a loose sense call a language. What we cannot expect, however, is that we can formalize the considerations that lead us to adjust our theory to fit the inflow of new information. No doubt we normally count the ability to shift ground appropriately as part of what we call 'knowing the language'. But in this sense, there is no saying what someone must know who knows the language; for intuition, luck, and skill must play as essential a role here as in devising a new theory in any field; and taste and sympathy a larger role.

In conclusion, then, I want to urge that linguistic communication does not require, though it very often makes use of, rule-governed repetition; and in that case, convention does not help explain what is basic to linguistic communication, though it may describe a usual, though contingent, feature.

I have a final reflection. I have not argued here, though I have elsewhere, that we cannot confidently ascribe beliefs and desires and intentions to a creature that cannot use language.[14] Beliefs, desires, and intentions are a condition of language, but language is also a condition for them. On the other hand, being able to attribute beliefs and desires to a creature is certainly a condition of sharing a convention with that creature; while, if I am right in what I have said in this paper, convention is not a condition of language. I suggest, then, that philosophers who make convention a necessary element in language have the matter backwards. The truth is rather that language is a condition for having conventions.

[14] See Essay 11.

Bibliographical References

Allen, Woody, 'Condemned', *New Yorker* (21 November 1977), 59.

Arendt, H., *The Life of the Mind*. Harcourt, Brace, Jovanovich, New York (1978).

Austin, J. L., and Strawson, P. F., 'Symposium on "Truth"', in *Proceedings of the Aristotelian Society*, Supplementary Volume 24 (1950).

Austin, J. L., *How to Do Things with Words*. Harvard University Press, Cambridge, Mass. (1962).

Bain, A., *English Composition and Rhetoric*. D. Appleton, New York (1867).

Bar-Hillel, Y., 'Logical Syntax and Semantics', *Language*, 30 (1954), 230–7.

Bar-Hillel, Y., 'Remarks on Carnap's Logical Syntax of Language', in *The Philosophy of Rudolf Carnap*, ed. P. A. Schilpp. Open Court, La Salle, Illinois (1963).

Bar-Hillel, Y., *Language and Information*. Addison-Wesley, Jerusalem (1964).

Barfield, O., 'Poetic Diction and Legal Fiction', in *The Importance of Language*, ed. M. Black. Prentice-Hall, Englewood Cliffs, New Jersey (1962).

Benacerraf, P., 'Mathematical Truth', *Journal of Philosophy*, 70 (1973), 661–79.

Beth, E. W., 'Carnap's Views on the Advantages of Constructed Systems Over Natural Languages in the Philosophy of Science', in *The Philosophy of Rudolf Carnap*, ed. P. A. Schilpp. Open Court, La Salle, Illinois (1963).

Black, M., 'The Semantic Definition of Truth', *Analysis*, 8 (1948), 49–63.

Black, M., 'Metaphor', in *Models and Metaphors.* Cornell University Press, Ithaca, New York, (1962).

Black, M., 'Meaning and Intention: An Examination of Grice's Views', in *New Literary History*, 4 (1972–3), 257–79.

Black, M., 'How Metaphors Work: A Reply to Donald Davidson', *Critical Inquiry*, 6 (1979), 131–43.

Bohnert, H., 'The Semiotic Status of Commands', *Philosophy of Science*, 12 (1945), 302–15.

Burge, T., 'Reference and Proper Names', *Journal of Philosophy*, 70 (1973), 425–39.

Burge, T., 'Reasoning about Reasoning', *Philosophia*, 8 (1979), 651–6.

Carnap, R., *The Logical Syntax of Language*. Routledge and Kegan Paul, London (1937).

Carnap, R., *Meaning and Necessity*. University of Chicago Press, Chicago. First edition (1947), enlarged edition (1956).

Cartwright, R., 'Propositions', in *Analytical Philosophy*, ed. R. J. Butler. Blackwell, Oxford (1962).

Cavell, S., 'Aesthetic Problems of Modern Philosophy', in *Must We Mean What We Say?* Charles Scribner, New York (1969).

Chomsky, N., *Aspects of the Theory of Syntax.* M.I.T. Press, Cambridge, Mass. (1965).

Chomsky, N., 'Topics in the Theory of Generative Grammar', in *Current Trends in Linguistics, 3*, ed. T. A. Sebeok. The Hague, Mouton (1966).

Chomsky, N., *Problems of Knowledge and Freedom*. Pantheon, New York (1971).

Church, A., 'On Carnap's Analysis of Statements of Assertion and Belief', *Analysis*, 10 (1950), 97–9.

Church, A., 'A Formulation of the Logic of Sense and Denotation', in *Structure, Method and Meaning: Essays in Honour of H. M. Sheffer*, ed. P. Henle, H. M. Kallen and S. K. Langer. Liberal Arts Press, New York (1951).

Church, A., 'Intensional Isomorphism and Identity of Belief', *Philosophical Studies*, 5 (1954), 65–73.

Church, A., *Introduction to Mathematical Logic*, Vol. 1. Princeton University Press, Princeton (1956).

Cohen, T., 'Figurative Speech and Figurative Acts', *Journal of Philosophy*, 72 (1975), 669–84.

Davidson, D., *Essays on Actions and Events*. Clarendon Press, Oxford (1980).

Davidson, D., 'Toward a Unified Theory of Meaning and Action', in *Grazer Philosophische Studien*, 2 (1980), 1–12.

Dummett, M., *Frege: Philosophy of Language*. Duckworth, London (1973).

Dummett, M., 'Truth', in *Truth and Other Enigmas*. Duckworth, London (1978).

Eliot, T. S., *Selected Poems*. Harcourt, Brace, Jovanovich, New York (1967).

Empson, W., *Some Versions of Pastoral*. Chatto and Windus, London (1935).

Feyerabend, P., 'Explanation, Reduction, and Empiricism', in *Scientific Explanation, Space and Time: Minnesota Studies in the Philosophy of Science*, 3. University of Minnesota Press, Minneapolis (1962).

Feyerabend, P., 'Problems of Empiricism', in *Beyond the Edge of Certainty*, ed. R. G. Colodny. Prentice-Hall, Englewood Cliffs, New Jersey (1965).

Field, H., 'Tarski's Theory of Truth', *Journal of Philosophy*, 69 (1972), 347–75.

Field, H., 'Quine and the Correspondence Theory', *Philosophical Review*, 83 (1974), 200–28.

Field, H., 'Conventionalism and Instrumentalism in Semantics', *Noûs*, 9 (1975), 375–406.

Foster, J., 'Meaning and Truth Theory', in *Truth and Meaning*, ed. G. Evans and J. McDowell. Clarendon Press, Oxford (1976).

Frege, G., 'On Sense and Reference', in *Philosophical Writings*, ed. M. Black and P. T. Geach. Blackwell, Oxford (1962), 56–78.

Geach, P. T., *Mental Acts*. Routledge and Kegan Paul, London (1957).

Geach, P. T., 'Assertion', *Philosophical Review*, 74 (1965), 449–65.

Geach, P. T., 'Quotation and Quantification', in *Logic Matters*. Blackwell, Oxford (1972).

Goodman, N., *Languages of Art*. Bobbs-Merrill, Indianapolis (1968).

Goodman, N., 'Metaphor as Moonlighting', *Critical Inquiry*, 6 (1979), 125–30.

Harman, G., 'Logical Form', in *Foundations of Language*, 9 (1972), 38–65.

Harman, G., 'Meaning and Semantics', in *Semantics and Philosophy*, ed. M. I. Munitz, and P. K. Unger. New York University press, New York (1974).

Harman, G., 'Moral Relativism Defended', *Philosophical Review*, 84 (1975), 3–22.

Henle, P., 'Metaphor', in *Language, Thought and Culture*, ed. P. Henle. University of Michigan Press, Ann Arbor, Michigan (1958).

Hintikka, J., *Knowledge and Belief*. Cornell University Press, Ithaca (1962).

Jackson, H., *The Eighteen-Nineties*. Alfred Knopf, New York (1922).

Jeffrey, R., review of *Logic, Methodology, and the Philosophy of Science*, ed. E. Nagel, P. Suppes, and A. Tarski, *Journal of Philosophy*, 61 (1964), 79–88.

Jeffrey, R., *The Logic of Decision*. McGraw-Hill, New York (1965).

Kripke, S., 'Is There a Problem about Substitutional Quantification?', in *Truth and Meaning*, ed. G. Evans and J. McDowell. Clarendon Press, Oxford (1976).

Kuhn, T. S., *The Structure of Scientific Revolutions*. University of Chicago Press, Chicago (1962).

Kuhn, T. S., 'Reflections on my Critics', in *Criticism and the Growth of Knowledge*, ed. I. Lakatos and A. Musgrave. Cambridge University Press, Cambridge, England (1970).

Lewis, D., 'General Semantics', in *Semantics for Natural Language*, ed. D. Davidson and G. Harman. D. Reidel, Dordrecht-Holland (1972).

Lewis, D., 'Radical Interpretation', *Synthèse*, 27 (1974), 331–44.

Lewis, D., 'Languages and Language', in *Language, Mind and Knowledge*, ed. K. Gunderson. University of Minnesota Press, Minneapolis (1975).

McCarrell, N. S., and Verbrugge, R. R., 'Metaphoric Comprehension: Studies in Reminding and Resembling', *Cognitive Psychology*, 9 (1977), 494–533.

Malcolm, N., 'Thoughtless Brutes', in *Proceedings and Addresses of the American Philosophical Association*, (1972–3).

Mates, B., 'Synonymity', in *Semantics and the Philosophy of Language*, ed. L. Linsky. University of Illinois Press, Urbana, Illinois (1952).

Mates, B., *Elementary Logic*. Clarendon Press, Oxford (1965).

Murray, J. Middleton, *Countries of the Mind.* Collins, London (1922).

Onions, C. T., *An Advanced English Syntax.* Routledge and Kegan Paul, London (1965).

The Oxford English Dictionary, ed. J. A. H. Murray *et al.* Clarendon Press, Oxford (1933).

Parsons, K. P., 'Ambiguity and the Theory of Truth', *Noûs*, 7 (1973), 379–94.

Prior, A. N., *Past, Present and Future.* Clarendon Press, Oxford (1967).

Putnam, H., 'The Meaning of "Meaning"', in *Mind, Language and Reality.* Cambridge University Press, Cambridge, England (1975).

Quine, W. V., *Mathematical Logic.* Harvard University Press, Cambridge, Mass. (1940).

Quine, W. V., *Methods of Logic.* Holt, New York (1950).

Quine, W. V., *Word and Object.* M.I.T. Press, Cambridge, Mass. (1960).

Quine, W. V., *From a Logical Point of View.* Second edition, Harvard University Press, Cambridge, Mass. (1961).

Quine, W. V., 'Two Dogmas of Empiricism', in *From a Logical Point of View.* Second edition, Harvard University Press, Cambridge, Mass. (1961).

Quine, W. V., 'On an Application of Tarski's Theory of Truth', in *Selected Logic Papers.* Random House, New York (1966).

Quine, W. V., 'Truth by Convention', in *The Ways of Paradox.* Random House, New York (1966), 70–99.

Quine, W. V., 'Ontological Relativity', in *Ontological Relativity and Other Essays.* Columbia University Press, New York (1969).

Quine, W. V., 'Reply to "On Saying That"', in *Words and Objections*, ed. D. Davidson and G. Harman. D. Reidel, Dordrecht-Holland (1969).

Quine, W. V., 'Speaking of Objects', in *Ontological Relativity and Other Essays.* Columbia University Press, New York (1969).

Quine, W. V., 'Comments on "Belief and the Basis of Meaning"', *Synthèse*, 27 (1974), 325–9.

Quine, W. V., *The Roots of Reference.* Open Court, La Salle, Illinois (1974).

Quine, W. V., 'On Empirically Equivalent Systems of the World', *Erkenntnis*, 9 (1975), 313–28.

Quine, W. V., 'Comments on "Moods and Performances"', in

Meaning and Use, ed. A. Margalit. D. Reidel, Dordrecht-Holland (1979).

Quine, W. V., 'On the Very Idea of a Third Dogma', in *Theories and Things*. Harvard University Press, Cambridge, Mass. (1981).

Quine, W. V., 'Replies to the Eleven Essays', *Southwestern Journal of Philosophy*, 10 (1981).

Ramsey, F. P., 'Facts and Propositions', reprinted in *Foundations of Mathematics*. Humanities Press, New York (1950).

Ramsey, F. P., 'Truth and Probability', reprinted in *Foundations of Mathematics*. Humanities Press, New York (1950).

Reichenbach, H., *Elements of Symbolic Logic*. Macmillan, London (1947).

Ross, J. R., 'Metalinguistic Anaphora', *Linguistic Inquiry*, 1 (1970). 273.

Scheffler, I., 'An Inscriptional Approach to Indirect Quotation', *Analysis*, 10 (1954), 83–90.

Scheffler, I., *The Anatomy of Inquiry*. Alfred Knopf, New York (1963).

Sellars, W., 'Putnam on Synonymity and Belief', *Analysis*, 15 (1955), 117–20.

Sellars, W., 'Truth and "Correspondence"', *Journal of Philosophy*, 59 (1962), 29–56.

Sellars, W., 'Conceptual Change', in *Conceptual Change*, ed. G. Pearce and P. Maynard. D. Reidel, Dordrecht-Holland (1973).

Strawson, P. F., and Austin, J. L., 'Symposium on "Truth"', in *Proceedings of the Aristotelian Society*, Supplementary Volume 24 (1950).

Strawson, P. F., 'Singular Terms, Ontology and Identity', *Mind*, 65 (1956), 433–54.

Strawson, P. F., *Individuals*. Methuen, London (1959).

Strawson, P. F., 'Truth: A Reconsideration of Austin's Views', *Philosophical Quarterly*, 15 (1965), 289–301.

Strawson, P. F., *The Bounds of Sense*. Methuen, London (1966).

Tarski, A., 'The Semantic Conception of Truth', *Philosophy and Phenomenological Research*, 4 (1944), 341–75.

Tarski, A., 'The Concept of Truth in Formalized Languages', in *Logic, Semantics, Metamathematics*. Clarendon Press, Oxford (1956).

Tarski, A., 'Truth and Proof', *Scientific American*, 220 (1967), 63–77.

Tharp, L., 'Truth, Quantification, and Abstract Objects', *Noûs*, 5 (1971), 363–72.

Thomson, J. F., 'Truth-bearers and the Trouble about Propositions', *Journal of Philosophy*, 66 (1969), 737–47.

Verbrugge, R. R., and McCarrell, N. S., 'Metaphoric Comprehension: Studies in Reminding and Resembling', *Cognitive Psychology*, 9 (1977), 494–533.

Wallace, J., *Philosophical Grammar*, Ph.D. Thesis, Stanford University (1964).

Wallace, J., 'Propositional Attitudes and Identity', *Journal of Philosophy*, 66 (1969), 145–52.

Wallace, J., 'On the Frame of Reference', *Synthèse*, 22 (1970), 61–94.

Wallace, J., 'Convention T and Substitutional Quantification', *Noûs*, 5 (1971), 199–211.

Wallace, J., 'Positive, Comparative, Superlative', *Journal of Philosophy*, 69 (1972), 773–82.

Wallace, J., 'Nonstandard Theories of Truth', in *The Logic of Grammar*, ed. D. Davidson and G. Harman. Dickenson Publishing Co., Belmont, California (1975).

Wallace, J., 'Only in the Context of a Sentence do Words Have Any Meaning', *Midwest Studies in Philosophy, 2: Studies in the Philosophy of Language*, ed. P. A. French, T. E. Uehling, Jr., and H. K. Wettstein. University of Minnesota Press, Morris (1977).

Weinstein, S., 'Truth and Demonstratives', *Noûs*, 8 (1974), 179–84.

Whorf, B. L., 'The Punctual and Segmentative Aspects of Verbs in Hopi', in *Language, Thought and Reality: Selected Writings of Benjamin Lee Whorf*, ed. J. B. Carroll. The Technology Press of Massachusetts Institute of Technology, Cambridge, Mass. (1956).

Wilson, N. L., 'Substances Without Substrata', *Review of Metaphysics*, 12 (1959), 521–39.

Index